Central Issues in Contemporary Economic Theory and Policy

General Editor: **Gustavo Piga**, *Managing Editor, Rivista di Politica Economica, Rome, Italy*

Published titles include:

Mario Baldassarri, John McCallum and Robert A. Mundell (*editors*)
DEBT, DEFICIT AND ECONOMIC PERFORMANCE

Mario Baldassarri (*editor*)
HOW TO REDUCE UNEMPLOYMENT IN EUROPE

Mario Baldassarri (*editor*)
THE NEW WELFARE
Unemployment and Social Security in Europe

Mario Baldassarri, Michele Bagella and Luigi Paganetto (*editors*)
FINANCIAL MARKETS
Imperfect Information and Risk Management

Mario Baldassarri and Bruno Chiarini (*editors*)
STUDIES IN LABOUR MARKETS AND INDUSTRIAL RELATIONS

Mario Baldassarri and Pierluigi Ciocca (*editors*)
ROOTS OF THE ITALIAN SCHOOL OF ECONOMICS AND FINANCE
From Ferrara (1857) to Einaudi (1944) (three volumes)

Mario Baldassarri, Cesare Imbriani and Dominick Salvatore (*editors*)
THE INTERNATIONAL SYSTEM BETWEEN NEW INTEGRATION AND
NEO-PROTECTIONISM

Mario Baldassarri and Luca Lambertini (*editors*)
ANTITRUST, REGULATION AND COMPETITION

Mario Baldassarri, Luigi Paganetto and Edmund S. Phelps (*editors*)
EQUITY, EFFICIENCY AND GROWTH
The Future of the Welfare State

Mario Baldassarri, Luigi Paganetto and Edmund S. Phelps (*editors*)
THE 1990s SLUMP
Causes and Cures

Mario Baldassarri, Luigi Paganetto and Edmund S. Phelps (*editors*)
INTERNATIONAL DIFFERENCES IN GROWTH RATES
Market Globalization and Economic Areas

Geoffrey Brennan (*editor*)
COERCIVE POWER AND ITS ALLOCATION IN THE EMERGENT EUROPE

Laura Castellucci and Anil Markandya
ENVIRONMENTAL TAXES AND FISCAL REFORM

Bruno Chiarini and Paolo Malanima
FROM MALTHUS' STAGNATION TO SUSTAINED GROWTH

Guido Cozzi and Roberto Cellini (*editors*)
INTELLECTUAL PROPERTY, COMPETITION, AND GROWTH

Debora Di Gioacchino, Sergio Ginebri and Laura Sabani (*editors*)
THE ROLE OF ORGANIZED INTEREST GROUPS IN POLICY MAKING

Luca Lambertini (*editor*)
FIRMS' OBJECTIVES AND INTERNAL ORGANISATION IN A GLOBAL ECONOMY
Positive and Normative Analysis

Riccardo Leoni and Giuseppe Usai (*editors*)
ORGANIZATIONS TODAY

Marco Malgarini and Gustavo Piga (*editors*)
CAPITAL ACCUMULATION, PRODUCTIVITY AND GROWTH
Monitoring Italy 2005

Gustavo Piga and Khi V. Thai
THE ECONOMICS OF PUBLIC PROCUREMENT

Central Issues in Contemporary Economic Theory and Policy
Series Standing Order ISBN 978–0–333–71464–5
(*outside North America only*)

You can receive future titles in this series as they are published by placing a standing order. Please contact your bookseller or, in case of difficulty, write to us at the address below with your name and address, the title of the series and the ISBN quoted above.

Customer Services Department, Macmillan Distribution Ltd, Houndmills, Basingstoke, Hampshire RG21 6XS, England

Environmental Taxes and Fiscal Reform

Edited by

Laura Castellucci
University of Rome 'Tor Vergata', Rome, Italy

and

Anil Markandya
Basque Centre for Climate Change, Bilbao, Spain

First published 2013 by
PALGRAVE MACMILLAN

Palgrave Macmillan in the UK is an imprint of Macmillan Publishers Limited, registered in England, company number 785998, of Houndmills, Basingstoke, Hampshire RG21 6XS.

Palgrave Macmillan in the US is a division of St Martin's Press LLC, 175 Fifth Avenue, New York, NY 10010.

Palgrave Macmillan is the global academic imprint of the above companies and has companies and representatives throughout the world.

Palgrave® and Macmillan® are registered trademarks in the United States, the United Kingdom, Europe and other countries.

ISBN: 978–0–230–39239–7

This book is printed on paper suitable for recycling and made from fully managed and sustained forest sources. Logging, pulping and manufacturing processes are expected to conform to the environmental regulations of the country of origin.

A catalogue record for this book is available from the British Library.

Library of Congress Cataloging-in-Publication Data

Environmental taxes and fiscal reform / edited by Laura Castellucci, Anil Markandya.
    p. cm.
    ISBN 978–0–230–39239–7
    1. Environmental impact charges. 2. Fiscal policy—Environmental aspects. 3. Taxation—Environmental aspects. I. Castellucci, Laura. II. Markandya, Anil, 1945–

HJ5316.E58165 2012
336.2′783337—dc23   2012012334

10 9 8 7 6 5 4 3 2 1
22 21 20 19 18 17 16 15 14 13

Printed and bound in Great Britain by
CPI Antony Rowe, Chippenham and Eastbourne

Contents*

---

* These chapters were originally published as articles in the journal *Rivista di Politica Economica* and have been reproduced exactly and in their entirety, including any minor typographic errors or omissions which may have been present in the original.

Environmental Taxes
and Fiscal Reform

Laura Castellucci*   -   **Anil Markandya**

University of Rome "Tor Vergata"   Basque Centre for Climate Change BC³,
Spain and University of Bath

The ordinary way of financing government expenditures is through taxation. This is not to say that borrowing is an extraordinary way of financing public expenditures, nor it is to deny other sources of financing such as charges to the public, rents, profits from the nationalized industries and sale of assets. It is to stress the fact that the "basics" of the budget-making process consists of different types of expenditures and revenues from taxation. Again it is not the idea of balancing the budget but rather the fact that the decision of how much to spend must be (loosely) linked to how much the government can reasonably expect to raise from tax collection without inducing more tax evasion and tax elusion. In other words, the decision on the level of expenditures has to be guided by the tax capability of a country. Tax effort and tax revenue are fundamental concepts that have been overlooked in the recent years, at least in Italy, and they must be resurrected. Although the prolonged international economic crisis requires a balance of both prudence in controlling expenditures and efficiency in tax collecting, the situation of public accounts is a serious one in several countries, including Italy. Fortunately, due to European Union obligations and especially to the adoption of the European currency[1], Italy has to strengthen both its tax effort and

---

* <laura.castellucci@uniroma2.it>; <Anil.markandya@bc3research.org>.
[1] The introduction of the Euro was a great achievement in terms of the European cohesion and the participation of Italy was very important for the

its tax revenue. In this regard there may well be a role for environmental taxation, which could serve the twin targets of internalizing pollution effects and raising revenue.

The best fiscal system for a country is the one that allows maximum revenue in the long run (and at least administrative costs) satisfying both efficiency and equity. To make environmental taxes to play a greater role in the fiscal system is totally consistent with these criteria. For, on one hand it is fair to tax "bads", such as pollution or excessive use of natural resources, rather than "goods", such as labor and profits, and on the other it is a known result in the received literature that green taxes have a positive effect on the development and diffusion of new technologies. GDP (and possibly employment) increases will follow, which in turn secure an increase in the long run tax revenue. Moreover since every fiscal system has adapted to the structure of the economy in which it operates, it is somehow contradictory to have in operation fiscal systems whose structure have been basically the same for the last 50/60 years while the economies have undergone great changes and transformation. Fifty years ago pollution problems, for example, were much less serious than they are today, when they pose serious threats to human health not to mention the rights of future generations. As it is another well established result in the literature, it is impossible to have a non distortionary fiscal system in practice; thus why not use the non-distortionary effects of environmental taxes to move away from consumption/production patterns which are harmful for the wellbeing of present and future generation? Some might say that green taxation has a regressive impact; this extremely important issue for the welfare of a society has been little investigated up to now, in the sense that the regressive impact has been more asserted than proved. This topic needs larger and deeper empirical investigation because vested interest may hide behind the superficially addressed argument, rather than a concern for the low income households.

According to many official governments' declarations, as well

---

country, as the present economic crisis has clearly shown. It is also definitely true that for Italy to adopt a restrictive budgetary policy as European constraint is of great help.

some treaties voluntarily signed by the international community of developed countries (and sometimes also less developed ones), two targets seem clear for the present century: to switch to a "low carbon society" and to go ahead with the Information and Communication Technology (ICT) economy. The state of the art suggests major investments are necessary to try to accomplish these results. Given the budget situation of the great majority of countries, these investments can only come from the private sector; but to deliver the required scale of investment the private sector needs the appropriate "signals" from the government which can take the form of tax savings from investments in low carbon alternatives or higher tax liabilities from investment in polluting technologies. Interestingly, in countries where opinion polls have been produced by association of firms themselves, such as in UK, it has emerged that the business sector would welcome the introduction of green taxes since the definition of the tax base, tax rates, exemptions *etc.*, would create the climate of certainty necessary to take decision about investments. The uncertainty about public sector endorsement of a precise environmental policy is deemed the greatest obstacle to private investment.

It seems that the time for a green tax reform has come but while the literature is already well equipped with theoretical results, more empirical studies are necessary, especially in some countries, such as Italy. To conduct such studies specific data which are not available at present are necessary. This fact comes as a confirmation that many changes in the economy, namely those traceable to the pollution effects of consumption and production patterns, have been overlooked and therefore not even caught by the statistical offices.

This special issue aims to encourage the study of fiscal reforms – consistent with the ongoing environmental problems. Disseminating concepts and results and addressing a variety of issues under different perspectives and in relation to different countries, has been the focus of the papers included here.

To provide the reader with a roadmap to the special issue, we have grouped the eight papers under three main titles. Besides an opening section devoted to the general issues, we choose to have one dedicated to the distributional effects as we see the topic of distribution as extremely important in any democratic society but rather overlooked in recent years, and another one dedicated to carbon taxation which is, by and large, the most (superficially) known argument when it comes to environmental policy.

In the opening paper Markandya, after having clarified the meaning and foundation of the green taxation, discusses the main theoretical argument on which most of the advocacy of green taxation rests today, namely the double dividend thesis. He then provides a comprehensive picture of types and results of empirical studies carried out so far. The second paper provides an overview on a less known topic: that of the Central and Eastern European countries (CEE). It shows how difficult it is to correct for environmental externalities. Even when a green tax is introduced it will more likely produce revenue rather than cure the external effects.

The next two empirical papers refer to very different countries and address the distributional issue from a different perspective: as an impact of environmental fiscal reforms in the first one and as environmental consequences of changes in income distribution the second one. While Sterner-Slunge paper is concerned with some African countries, the Castellucci-D'Amato-Zoli looks at distributional issues in Italy. Since African countries are not the most studied countries in the literature, the results of the paper are important in themselves besides being very interesting in that they show that the regressivity impact of energy taxation is not generally found and in many cases it may well be (slightly) progressive. An analogous lesson can be learnt from the Italian case in the sense that while a redistribution towards lower income households may increase pollution (deteriorate environment) it might also decrease pollution. It depends on the emission intensity of consumption. If such intensity is assumed to be constant, the environment deteriorates as a consequence of a progressive redistribution but if such intensity varies in a plausible manner the results are reversed. To know the emission intensity of consumption is an empirical question which is at present unsolved due to the lack of data.

Attitudes to green or environmental taxation have been and still are at least *skeptical*, notwithstanding the received theoretical incontrovertible literature showing such taxes as the cost minimizing instrument. This issue is explored in the four remaining papers. Two of them are country and sector specific while two are general. Here again we underline the fact that empirical researchers on Italy and on sectors, such as aviation and also manufacturing, are not that abundant; this is why we think the two papers may help to enrich the debate while providing some lessons. Finally, the never ending search for the optimal policy cannot go without an up to date survey of the literature (Galarraga-Ansuategi) and our special issue on tax reform cannot go without addressing the question of tax harmonization (D'Amato-Spisto). Without claiming that this third section surveys all aspects of the present state of carbon taxation (which is huge), the coverage is broad and able to give a good snapshot of the topic.

I - GENERAL ISSUES

Environmental Taxation: What Have We Learnt in the Last 30 Years?

Anil Markandya*

Basque Centre for Climate Change BC³ and University of Bath

The paper surveys the literature on environmental taxation in the context of the climate change problem. It reviews the use of taxes as instruments of environmental management and notes the strengths and weaknesses of the applications. The performance of taxes relative to permits depends on the presence of uncertainty whose impacts are explored. The paper also reports on simulation studies concerned with double dividend implications of a carbon tax combined with a reduction in other taxes. The models for the EU find that a switch in taxation from labour to carbon/energy would increase employment, reduce carbon emissions and increase GDP. [JEL Codes: H21, H23, Q54, Q58]

Keywords: taxation; environment; double dividend; carbon taxes.

1. - Introduction

There is no formal definition of "Environmental Taxation", although most people have a similar understanding of what they mean when they use the term. Taxes are considered "environmental" if, in some way, they promote the protection of the environment and the natural resources of the planet. Thus any tax that reduces the use of fossil fuels could be considered green, and in

---

* <*anil.markandya@bc3research.org*>. I would like to thank Professor Castellucci for comments, colleagues at Bath University, especially Professor Heady for joint work on which some sections of the paper are based, and Professor Smith from UCL for comments on a version that was presented in Copenhagen. All errors that remain are my own.

this context the history of taxation of petroleum products goes back a long way (Denmark for example had a petrol tax in 1917). Other green taxes that have been in existence for a long time include taxes on coal, and on the exploitation of natural resources[1]. In Russia in the Tsarist period of the 20th century, for example, around 90 percent of local revenues came from taxes on the use of natural resources.

So, in this sense environmental taxes are not new and have been a part of the structure of taxation well before the Green Movement was even thought of. What is new, however, is the targeting of such taxes to meet specific environmental objectives. The earlier taxation of oil and natural resources was seen largely as a means of collecting revenues in a way that was not too painful. Now we see them also as a possible instrument for reducing environmental burdens. The level of such taxes has, of course, to balance the benefits in terms of environmental gains against the costs of taxation, in terms of reduced consumption and production. Back in the 1930s economists established the framework for calculating the "optimal" tax as a balance between these two forces. The key concept in this analysis was that of negative "externalities". Anthropogenic activities were said to create negative externalities when the actions of one person or group resulted in damages to another group, and when the first group did not take proper account of such damages. The obvious example is a polluter who does not take account of the consequences of his emissions on others when deciding on the level of his own activities.

More recently proponents of green taxation have stressed another potential of such taxation, and that is the possibility of shifting the tax base away from the taxation of labour, capital and goods and services, to the taxation of pollutants. The claim is that such a shift creates benefits in terms of a more efficient fiscal system (with lower welfare losses from taxation) as well as, possibly, stimulating employment in countries where there is structural unemployment (Pearce, 1991).

---

[1] To be sure there have also been substantial subsidies on natural resource use as well. For a discussion of subsidies see von Moltke A. *et al.* (2004).

Thus there are two strands in the green taxation literature: the first that seeks to determine such taxes in terms of the environmental benefits at the "micro" level, and the second that seeks to justify them in terms of broader fiscal and employment benefits. While the general case for green taxation is made on the grounds cited above, critics point to a number of problems. First some question the link between the theory of internalizing externalities and the implementation of such a system in practice. Even the best practical tax design would not meet the assumptions under which such a tax could be guaranteed to be welfare maximizing[2]. Second and related to that, there are economists who argue in favour of other instruments, such as permits, standards, subsidies *etc.*, as more efficient tools for environmental regulation. Even these, however, do not in practice meet the conditions for welfare optimality. Both these debates take us into the realm of the "second best", where we have to compare alternative policies (*e.g.* using environmental taxes *versus* the use of direct controls) to achieve the desired objectives. Third, green taxation has been criticized for its possible negative impact on competitiveness, employment and growth. Ironically then there are those who suggest that green taxation can increase employment and even enhance growth and those who claim the opposite.

This paper addresses these questions in turn with a focus mainly on taxes that address climate change. It begins by setting out the case for Environmental Taxation on externality grounds and then on macro-fiscal grounds. This is followed by a review of proposed climate taxation in Europe and worldwide. Finally the paper concludes with a discussion of the open questions and issues for debate.

---

[2] A tax is said to be welfare maximizing if it results in an allocation of resources such that no person can be made better off without someone else being made worse off.

2. - The Theoretical Arguments For and Against Green Taxation As a Tool for Environmental and Economic Policy

2.1 *The Pure Externality Argument*

With the onset of industrialization the range of external effects needing some action by the authorities grew considerably. The processes of industrial production involved large-scale use of fossil fuels, which generated harmful pollutants, as well as the use of chemicals and other inputs that created gaseous, liquid and solid waste. From the early 19th Century we see those responsible for environmental regulation struggling to find the best way of dealing with this problem. Until recently the measures taken to address this problem involved passing a law, or issuing an administrative order, proscribing certain practices and requiring others to be undertaken. In the UK, for example, factories were ordered, by various parliamentary acts passed between 1820 and 1926, to reduce the output of smoke, and more recently the burning of coal was banned in certain urban areas. As transport became a major source of pollution, the use of more polluting fuels, such as lead, was banned and vehicles were required to be fitted with devices that reduced emissions.

All these environmental measures are referred to as command and control regulations. The authorities tell you what you must or must not do, and there are penalties under civil and/or criminal law if you fail to comply. When economists started to address environmental concerns their instinctive reaction was to look for alternative methods of regulation that did not involve compulsion but that relied on economic incentives to achieve the same goals. The British economist Pigou first noted that if you could tax the activity generating a negative externality (*i.e.* one causing harm to third parties or to the environment), the party responsible would reduce the intensity of that activity. And by selecting the tax level suitably, the authorities could achieve whatever goal they wished in terms of reducing the negative external effects (Pigou, 1932).

Principles Behind Economic Instruments: Controlling Emissions to Air and Water

The objective in theory for the use of a tax or other economic instrument is to regulate the level of an externality generating activity to its optimal point. This point is defined as one where reduction in the additional damage caused by the activity is equal to the cost of abating that additional amount of the activity. Graph 1 shows how this is arrived at[3].

In the absence of any regulations the enterprise will generate an output of OD, and will undertake no abatement of emissions because it has no incentive to do so. As it undertakes reductions in emissions (by adopting clean technology, or by using end of pipe clean up or by reducing output, or any combination of these), it incurs costs. These costs are shown by the Marginal Cost of Abatement curve, which gives the additional costs incurred when emissions are reduced by one unit. The curve slopes from right to left because, as more and more reductions are undertaken, the additional, or marginal costs, rises (enterprises undertake the lower cost options first). At the same time, the emissions are known to cause damages. At the level of emissions OD, damages are OH as shown in the Graph, and as emissions decline, so do the marginal damages. The underlying assumption in the Graph is that these damages (to health, property, ecosystems *etc.*) can be measured in money terms. This is a controversial assumption and, indeed does not hold in all cases. But it is made here for the convenience of showing what, *in principle*, the optimal level of an externality would be.

As emissions are reduced there is a reduction in damages and an increase in abatement costs. For a small reduction of Δ, from OD, the addition costs and reduced damages are also shown in Graph 1. Clearly the damages fall by more than costs of abatement increase, (there is a net gain equal to the dark-shaded area) and so the reduction of Δ is justified. This holds for all reductions

---

[3] This Graph is meant to add our understanding of the issues involved in internalizing an externality. It is not a complete description of the problem and should not be seen as such.

GRAPH 1

THE OPTIMAL LEVEL OF CONTROL OF A POLLUTANT

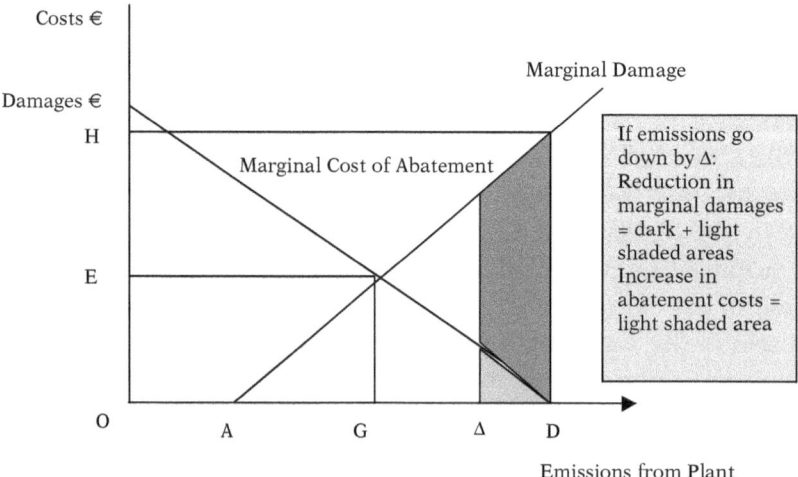

up to the point at which the marginal cost of abatement and marginal damage are equal – *i.e.* at the point where the emissions equal OG and the marginal costs and damages are both OE. *This is the optimal point for the externality.*

There are a number of important points to note about this analysis, which is the fundamental theoretical basis for determining the control and regulation of externalities.

A. The optimal regulation of emissions does not imply a level of zero emissions. Environmentalists would say that an ideal (*i.e.* optimal) situation would be with no emissions but economists point out that with a requirement of no allowed emissions from the activity the costs of abatement generally exceed the damages and this is not optimal. This is the major difference in approach between economic and ecological solutions.

B. The optimal solution can be obtained in a number of ways. These are:

i. An order to the polluter to produce only OG in the form of emissions, which is a direct or **COMMAND AND CONTROL** solution. This may seem easy and indeed would be so if there were only one plant generating the emissions. In practice the marginal cost of abatement curve is derived from the activities of lots of polluters, and the reductions along it are not undertaken at only one plant. Hence the amount of information needed to implement this command and control solution is very large and, as noted later, does not offer any flexibility in terms of response to the polluters.

ii. A TAX OR CHARGE of OE per unit of emissions. With such a charge the emitters will undertake reductions to the point OG because the costs of abatement are less than the charge.

iii. Instead of a charge, the polluter could be given a SUBSIDY for each unit of reduction equal to OE. This acts in the same way as a charge; the polluter finds it pays to make reductions to the point OG because the subsidy it receives exceeds the costs of abatement. Beyond that point the subsidy does not cover the costs of abatement.

iv. Issue TRADABLE PERMITS equal to OG and allow polluters to emit as much as they like, as long as they are in possession of a permit. It can be shown that in such a situation a market will develop for permits, with a price equal to OE. Those whose costs of abatement exceed OE will buy permits to cover those costs. The permits can be given to emitters in proportion to existing emissions or they can be auctioned. In the former case if a polluter was currently emitting 100 units out of a total of 1,000 and the authority wanted a overall reduction from 1,000 to 800 (*i.e.* 20 percent), it would allocate the polluter only 80 units, with the right to buy more if he needed them or less if he deemed it preferable to make a bigger reduction than 20 units.

v. Related to the last case, we could think of a MARKET FOR EMISSIONS RIGHTS. The laws would have to be passed defining these property rights for emissions, creating a registry of ownership of these rights and allowing them to be traded as desired. A full market solution is perhaps not easily implemented for emissions, but could be implemented for development rights on land. An owner of a parcel of land might have prior development rights on that land when he purchased it. For reasons of conservation these rights may be revoked on that land but, with a market in rights, he could demand the right to similar rights elsewhere.

Thus taxes are only one of the possible solutions to the externality problem and there is a great deal of debate as to which instrument is the most appropriate. There is fair agreement, however, on two things. The first is that the subsidy solution is inferior both to the tax or tradable permit/emissions rights solutions and the second is that frequently the market based options of taxes or permits result in a more efficient outcome than a command and control option. In Graph 1 it is difficult to see the benefits of the market-based options because all options result in the optimal solution of OG emissions. But this is misleading because in reality we do not have full knowledge of which polluter has which cost curve and we do not know the damage curve with much precision. Hence we cannot identify the optimal solution precisely. Sophisticated studies that take account of imperfect information show that in quite a wide set of cases, imposing a charge, which may be approximate, yields better results (in terms of total costs of abatement in achieving a given target reduction) than using command and control policies (Tietenberg, 1990).

The preference of taxes over subsidies as a means of getting to the optimal point is a consequence of a number of arguments. First, although subsidies "mirror" incentives to taxes in the simple diagram, in reality they are not complete mirrors because they also increase the profitability of the related activities and therefore can result in increased emissions on that count[4]. Second, there is the possibility of corruption when handing out subsidies. Third subsidies have to be paid from some source and usually that source is general taxation, which itself has a distortionary effect. The latter is measured as the "marginal cost of public funds", with a euro of taxes having a welfare cost of more than one euro (for the EU this marginal cost is around € 1.25-1.3 (Snow and Warren, 1996). Fourth there are difficulties in defining who should receive the subsidy. Would a payment be made to an enterprise that closed down?

---

[4] More generally the actual value of the MAC curve will depend on the choice of instrument and, with some command and control policies its shape may not even be monotonic.

If so, for how long would this payment be made? For all these reasons, economists have a strong preference for taxes over subsidies as the instrument of choice to correct externalities.

The second area where there is broad agreement is the benefits of tax scheme or market permit scheme over command and control on the grounds of the increased flexibility that they offer. Graph 1 does not show this, but, as we discuss in Section 3, there is a real benefit in allowing polluters to have some choice in how they reduce their emissions. The more strictly the actions are prescribed, the more likely it is that the chosen solution will be more costly than necessary.

The choice between taxes and permits partly depends on political issues of who is made to bear the cost of the adjustment, and partly on economic considerations of uncertainty. As Weizmann (1974) showed, if we are not certain of the costs of abatement and impose the wrong (too high) tax rate, the cost will be borne by industry and we will miss the emissions reduction target. On the other, hand if we use permits and are too strict in the number issued, we will over reach the emissions target, but at a possibly higher cost in terms of cuts in output and employment. The actual argument is quite sophisticated and depends on the slopes of the abatement and damage curves. In practice, given that empirical studies in a number of areas suggest that the marginal damage curve is relatively flat, a tax is more likely to get you the right answer than a permit scheme. We return to this issue in the next section in the context of climate change.

On the choice between permits and taxes, there is also the question of to whom permits are initially allocated. This issue, which creates both problems and opportunities, is discussed again in Section 3. We should note, however, that with small numbers of polluters, a permit scheme is less viable as the permit trading will take place in thin markets.

Environmental Taxes in Practice

In practice we have seen a significant increase in the use of fiscal instruments to address environmental problems. A range of pollution charges and some charges on the use of natural resources have been implemented in the EU15, as a national response to addressing environmental problems in an effective way, strongly supported by various EC declarations. Table 1 summarizes the situation as of 2001[5].

There were a wide range of instruments – 142 in total at that date and many more now – with the most common applications being in user charges for waste and water (all 15 countries have them). A major increase in the number of instruments took place in the 1990s. Motivations behind the instruments are essentially 3: *(a)* raising revenue for public environment and related activities, *(b)* providing an incentive to reduce emissions and/or save on the use of natural resources and *(c)* covering the costs of delivery of environment related services (*e.g.* waste water collection and treatment). The most notable features of the system of charges are the following.

Incentive Effects and Revenue Effects

A. The environmental effect of the taxes are estimated to be positive but small (with some exceptions). This has largely been due to low rates at which they are levied and the myriad exemptions that have been granted, on the basis of hardship, possible employment and competitiveness effects, *etc.*

B. The design of the taxes has given more emphasis to revenue raising than to the incentive effects (which would require much higher taxes in most cases). The revenues are often earmarked for specific environmental measures, so that the government address certain environmental problems.

C. Incentive effects of resource charges are limited because of the way the charge is levied. For example, if a user charge is levied on water, and is paid based on the size of the house, there is little incentive to reduce water consumption, as the amount paid does

---

[5] Excluded from the table are fuel and excise taxes and taxes on sulphur.

not depend on the level of consumption. Metering for water is still not widespread in the EU, making water charges a cost recovery instrument rather than an incentive based one. Charges based on amounts of wastewater generated are, however, somewhat more common, and some countries levy additional wastewater taxes, at rates that vary considerably across countries.

D. For waste the same problem arises as far as incentive effects are concerned; rarely are charges related to amounts of waste generated, although variable charging is being introduced by a few municipalities in Austria, the Benelux countries, France, Italy and Switzerland. There are taxes on waste disposal in 10 countries. These may have some incentive effects as the charges paid by the municipalities encourage them to recycle and find other ways to reduce the waste generated. Rates on landfill range from €3-€30 per ton. A full assessment of the impact of these taxes on amounts sent to landfill sites has not been carried out but earlier studies in the US on the "pay-by-the-bag" programs found significant reductions in amounts generated (OECD, 1993, Repetto *et al.*, 1992).

E. A handful of countries impose taxes on agricultural inputs – pesticides and fertilizers. Some, however, provide incentives for increased use by, for example, exempting them from VAT. Some countries that have imposed taxes on these inputs have seen a decline in their use (*e.g.* Netherlands, Denmark).

F. Taxes on products for environmental reasons are growing in popularity (most have been introduced since the late 1990s). The purpose is mainly to defray the costs of disposal of the products, including, in some cases, handling illegal disposal. Incentives to avoid improper disposal and to recycle are provides by "Take Back Schemes" and "Deposit Refund Schemes", which are used for batteries, disposable containers, lamp bulbs, refrigerators and some kinds of packaging. In earlier US studies these instruments have been found to reduce the amounts of waste and the costs of waste management significantly. The bottle bills in the US have reduced litter by 10 to 39 percent and solid waste by 1 to 6 percent; and kerbside collection programmes obtain recycling percentages of around 35 percent for glass containers and 25 to 56 percent for aluminium cans (Repetto *et al.*, 1992). As far as the impact of the producer-based recycling schemes is concerned, it is considered too early to make an evaluation (that was the case in 1993). Nevertheless it is encouraging to note that the German Green Dot scheme has obtained the participation of over 50 percent of households by 1993, and over 80 percent is expected by this year.

G. In terms of revenue generation, environmental taxes are still a minor part of total government taxes. As a percentage of total tax revenue they range from a low of 0.3% (Portugal) to a high of 5.9% (Netherlands). Energy taxes, on the other hand (which also have environmental impacts) are more significant – ranging from 3.2 to 8.4% of the total. Together the two taxes can add to as much as 10% of total tax revenue.

H. Overall, therefore, environmental taxes have not had strong incentive effects, but the case studies carried out showed that even a small tax can have a strong awareness effect, which is hard to measure, but which may be nevertheless, quite real.

2.2 *The Fiscal-Employment Benefits Argument*[6]

The case for the macroeconomic benefits of green taxation in conjunction with the environmental benefits is also referred to as the "double dividend"; the argument being that a switch to environmental taxation, combined with a reduction in labour or other taxes in a fiscally neutral way, would lead to a double dividend. The idea behind this suggestion is that environmental taxes not only produce improvements in the environment (the first dividend) but also generate substantial amounts of government revenue. This additional government revenue would allow governments to reduce the rates of other taxes in the economy while maintaining a constant level of total revenue and expenditure: the revenue-recycling effect. As these other taxes are generally regarded as distortionary (interfering with the efficient functioning of markets), the reduc-

---

[6] This section draws on MARKANDYA A. (2005), but updates that discussion in the light of subsequent developments.

tion in their rates can be seen as improving efficiency and thus producing a second benefit from the adoption of environmental taxes[7]. In the following discussion we could think of the tax that is being imposed as a carbon tax, and the taxes being removed as employment or other taxes. Indeed the following section looks explicitly at simulation studies of these combinations of taxes.

Gross Welfare Double Dividend vs. *Employment Double Dividend*

The literature in this area identifies two "second" dividends: a "gross welfare" dividend and an employment dividend. The *gross welfare dividend* arises because the tax changes reduce the distortions in consumer choice that result from sales and other taxes. The word "gross" indicates that one is not accounting for the welfare gains from an improved environment. The *employment dividend* arises because one possible distortionary effect of taxation is the reduction of employment. Such a reduction in employment could result from taxes that are obviously related to employment, such as income taxes and social security taxes, but also from taxes that affect the real value of workers' wages, such as value added taxes and excise duties. Thus one aspect of the double dividend could be an increase in employment that follows from a reduction in one or more of these taxes. As discussed later, the literature suggests that it is easier to obtain the employment dividend than the gross welfare dividend.

Although a large part of the theoretical double dividend literature deals with the gross welfare dividend, the focus of the policy discussion has been mainly on employment, at least in Europe. The literature on the gross welfare dividend generally does not address issues of employment: it assumes that there is no involuntary unemployment and places no particular value on addi-

---

[7] The rate of the tax on the environmentally damaging pollution is not equal to the Pigovian tax as described above, but rather equal to a weighted average of such a Pigovian tax and a revenue raising tax. This point has been made by a number of authors (BOVENBERG A.L. and VAN DER PLOEG F., 1994*b*, 1996; LIGTHART J.E. and VAN DER PLOEG F., 1999).

TABLE 1

ENVIRONMENTAL CHARGES AND SIMILAR INSTRUMENTS IN THE EU

Instrument	AU	BE	DK	FIN	FR	DE	EL	IRL	IT	L	NL	PO	ES	SW	UK	Total
Energy/Carbon Tax		X	X	X		X			X		X			X	X	8
Nox Charge				X	X								X[1]	X		4
Sulphur Tax			X	X										X		3
Agricultural Inputs																
Pesticides		X	X	X										X		4
Fertilizers	X[2]	X	X[3]	X[4]							X[5]					5
Eco Taxes on Goods																
Batteries	TBS	X	X			TBS			X					X		6
Plastic Carrier Bags			X			TBS										2
Disposable Containers	DRS	X	X			TBS			X					X		6
Tyres			X	X	X									X		3
CFCs/Halons			X													1
Disposable Cameras		X														1
Lubricant Oil			X	X					X				X	X		5
Oil Pollution				X					X							2
Others	DRS[6]	X[7]	X[8]		X[9]				X[10]		X[11]		X[12]	X[13]		8
Waste																
User Charge	X	X	X	X	X	X	X	X	X	X	X	X	X	X	X	15
Landfill Tax	X	X	X	X	X	X			X			X		X	X	10
Hazardous Waste Tax		X	X[14]	X[15]				X		X						5
Others			X		X[16]				X		X[17]					4
Water																
User Charge	X	X	X	X	X	X	X	X[18]	X	X	X	X	X	X	X	15
Water Tax/Abstraction Tax			X	X	X	X[19]			X		X[20]		X[21]			7
Waste Water Charge	X	X	X	X	X	X		X	X		X		X	X	X	12
Others			X	X[22]	X											3
Aggregates Tax		X[23]	X											X	X	4
Noise Charge		X	X	X	X	X			X		X		X	X	X[24]	9
Total Number	8	13	21	14	12	10	2	4	14	3	10	3	7	15	7	142

Source: ECOTEC *et al.*, 2001.

Notes: DRS = Deposit Refund Scheme; TBS = Take Back Scheme (*i.e.* supplier is obliged to make arrangements to take back the product after it has been used).
1. Only a regional tax - in Galacia; 2. Tax on growth promoter fertilizers only; 3. Tax has been abolished; 4. Tax is a mineral surplus tax; 5. Covers lamp bulbs, refrigerators and freezers and packaging; 6. Tax on packaging, surplus manure, heavy accidents, and ionizing radiation; 7. Tax has been abolished; 8. Tax on chlorinated solvents, disposable tableware, light bulbs, PVC and junk mail. DRS on reusable containers (beer and soft drinks); 9. Packaging tax, paper tax, tax on mines and tax on natural sites; 10. Packaging tax and tax on aggregates; 11. Surplus manure charge; 12. Eco-tax on tourism in Balearics; 13. Tax on gravel, limestone, packaging charge, vehicle scrapping charge; 14. Tax on electronic and electrical waste; 15. Nuclear waste management charge; 16. Tax on emissions from incinerators; 17. Waste charge on "disposal of white and brown good decree"; 18. Local taxes not related to water consumption; 19. Water abstraction charge (regional level); 20. Tax on groundwater only; 21. Also water sanitation charge and charge on spills to coastal waters; 22. Fish management charge; 23. Tax applies at regional level only (Flanders); 24. Air passenger duty.
Country Codes: AU-Austria; BE-Belgium; DK-Denmark; FIN-Finland; FR-France; DE-Germany; EL-Greece; IRL-Ireland; IT-Italy; L-Luxemburg; NL-Netherlands; PO-Portugal; ES-Spain; SW-Sweden; UK-United Kingdom.

tional employment creation. The idea is that there is very little gain in individual welfare in moving somebody from voluntary unemployment into employment, in contrast to the very substantial gains in moving somebody from involuntary unemployment into employment. Of course, both types of employment creation can improve tax revenue, and that effect is considered in the theoretical literature even when issues of employment are not addressed.

The pathway by which tax reductions might increase employment depends crucially on whether or not the labour market is in equilibrium, with demand equal to supply. If there is disequilibrium in the labour market, with supply greater than demand and consequent involuntary unemployment, employment creation requires an increase in labour demand. This could be achieved by reducing the cost of employing labour, for example by reducing employers' social security taxes. It is important to note that any increase in employment from this policy does not necessarily imply a reduction of unemployment by the same amount (or at all), because the increased availability of jobs may induce additional people to enter the labour force. The estimates of employment creation that are quoted later in this chapter should be interpreted with this point in mind.

On the other hand, if the labour market is in equilibrium, with demand equal to supply and no involuntary unemployment, an increase in employment requires an increase in labour supply. This could be achieved by increasing the returns to work, by reducing direct taxes on labour income or by reducing sales taxes on goods that workers wish to buy, provided that workers respond positively to such increased incentives. The part of the theoretical double dividend literature that has dealt explicitly with employment has concentrated mainly on the case of involuntary unemployment, not on this case.

Weak vs. *Strong Dividends*

The primary purpose of environmental taxes is to reduce damage to the environment by increasing the costs of harmful actions, such as the burning of fossil fuels that produces carbon dioxide.

The idea is that consumers and firms will then be forced to take account of the effects of their actions on the environment. For this to work properly, the size of the taxes should equal the monetary value of the marginal environmental damage that the actions cause. Taxes that meet this requirement are referred to as "Pigouvian" taxes, which we discussed in the previous section.

If the revenue from such Pigouvian environmental taxes were sufficiently large to fund all government expenditures, the existing distortionary taxes could be completely removed. Then the economy would be undistorted by either taxation or environmental externalities. The double dividend would be a reality in both welfare and employment terms.

However, most governments have expenditure levels that are more than 40 percent of their GDP, and Pigouvian taxes will not raise that level of revenue. It is therefore necessary to consider the effect of environmental taxes as *reducing* rather than entirely *replacing* other taxes. This means that the interaction between environmental taxes and other taxes – the tax-interaction effect – has to be considered, and it is this interaction that causes the analysis to be so complicated.

In order to understand this interaction, it is helpful to follow Goulder's (1995) distinction between the "weak double dividend" hypothesis and the "strong double dividend" hypothesis. The weak double dividend is simply concerned with what is done with the revenue from environmental taxes, saying that it is better to use this revenue to reduce the rates of distortionary taxes than to provide lump-sum payments to citizens. The strong double dividend says that the replacement of some existing taxes with environmental taxes will reduce the distortionary cost of raising the current level of government revenue, thus allowing real incomes and consumption to rise.

The weak double dividend has been shown to hold in almost all models. The most important exception is when the lump-sum payments are markedly better than tax reductions at raising the incomes of poor households.[8] However, the details of these results are not worth discussing here because the weak double dividend

---

[8] As this sentence illustrates, the theory can also take account of the

is simply about how to spend the environmental tax revenue. Although it may have some implications for how environmental taxes are spent and possibly on the choice between taxes and other instruments, it says nothing to enhance the case for implementing environmental taxes specifically to replace other taxes. It is the strong double dividend that needs to be true in order to claim that environmental taxes can contribute to the efficiency of the economy in other ways than improving the environment. The conditions for the existence of the strong double dividend require more sophisticated analysis, and there is a wider range of disagreement.

Because the strong double dividend is concerned with reducing the distortionary cost of the tax system, the analysis can only be fully understood in the context of the theory of optimal taxation: a theory that deals with the problem of minimizing the distortionary costs of a tax system that generates a given level of government revenue.

The first important fact to be aware of is that the theory of optimal taxation until recently has not concerned itself with environmental issues.[9] It is simply concerned with raising revenue efficiently, and so we can refer to such taxes as "revenue-optimal". Thus a revenue-optimal set of taxes is one that minimizes its effect, as measured by a "distortionary cost", on the actions of market participants, without regard to its environmental effect.[10] If a country has adopted a revenue-optimal set of taxes, there is no possible change to those taxes that will raise the same revenue at a smaller distortionary cost. In particular, the imposition of a higher rate of tax on a good that damages the environment cannot reduce the distortionary cost of the tax system, and can generally be expected to increase it. This implies that *the strong double div-*

---

distributional effects of taxes. However, these have not been paid much attention in the double dividend literature and so will not be emphasized here.

[9] SANDMO A. (1975) was an exception. More recently the work of PARRY I.W.H. *et al.* (*e.g.* PARRY I.W.H., 1998) has developed on that and begun to look at the environmental dimension more seriously.

[10] This concept of revenue-optimal taxes takes account of all the effects of tax changes, including those that result from the shifting of the tax burden between different groups in society.

idend cannot be true in an economy where the taxes are revenue-optimal.

Of course, this does not mean that there is never a strong double dividend, because it is unrealistic to suppose that countries currently have revenue-optimal taxes. What it does mean is that a strong double dividend exists when (and only when) the imposition of an environmental tax moves the tax structure closer to the revenue-optimum. Thus, those parts of the literature that claim to show the existence of a strong double dividend are based on presumed situations in which the existing taxes are not revenue-optimal and environmental taxes produce a move towards the revenue-optimum[11]. In contrast, the papers that show the absence of a strong double dividend assume either that taxes are already revenue-optimal or that the environmental tax does not move the system towards revenue-optimality. Which of these situations applies to any particular country is, of course, an empirical question, where computer simulation models are useful.

The Employment Double Dividend: The Case with Involuntary Unemployment

In economic terms unemployment is caused by the wage being higher than its market clearing value. This leads to a situation where the demand for labour is less than its supply, and the result is involuntary unemployment. In this situation, the only way to create additional employment is to increase the demand for labour. This section analyses how the use of environmental taxes to replace in part existing taxes might achieve such a demand increase.

There are several possible explanations for the "high" wage, including trades unions and various models of asymmetric information, but the analysis is easier if we start by simply taking the

---

[11] A previous commentator has rightly noted that, while true, this is not particularly useful in terms of the design of the environmental tax from the environmental perspective.

real after-tax wage as fixed[12] and only later look at the implications of how it is determined. Taking this approach, the important aspect of the unemployed economy is the distortion of the wage, and the standard response from optimal tax theory is to look for ways to reduce that distortion either by direct subsidy or by introducing offsetting taxes elsewhere.

It might be thought that the distortion of the wage could be reduced directly by reducing social security (payroll)[13] taxes, while imposing environmental taxes to replace the lost revenue. However, it must be recognized that the environmental taxes will increase the cost of goods that workers buy, thus tending to reduce the REAL wage. This is the tax-interaction effect. It implies that workers will demand wage increases to restore the previous value of the real wage and this will offset the effects of the reduced payroll taxes. In other words, the move from payroll taxes to environmental taxes has not reduced the taxation of workers; it has simply rearranged it.

In order for a move from payroll taxes to environmental taxes to increase employment, the taxation of workers must be reduced. This can be achieved in two possible ways: *(i)* the shifting of the tax burden from workers to other groups, and *(ii)* improvement in the efficiency of the tax system. We deal with these two possibilities in turn.

Shifting the Tax Burden

One case in which a shift from payroll taxes to environmental taxes could increase employment is when some consumers are not workers. For example, imagine that some consumers live entirely (or much more than average) on capital income. The imposition of a sales tax on an environmentally damaging good that

---

[12] This is based on the idea that workers are interested in what they can buy with their (after-tax) earnings, and that they can enforce a particular minimum level of this.

[13] In principle, cutting income tax would also reduce labour costs by reducing the before-tax wage rates that workers demand.

was used to reduce labour taxes would move some of the tax burden from workers to non-workers, and so reduce the distortion of the labour market (provided that the non-workers did not emigrate as a result). In other words, the revenue raised from the sales taxes would be more than sufficient to generate income to the government that allowed it to reduces the cost of labour to employers, and so encourage employment. This would create the employment double dividend, but at a cost to non-working consumers. The employment double dividend has arisen by shifting the tax burden from workers to non-workers.

It is worth noting that the same effect can occur with two other groups of non-workers. The first are people on state benefits, but only provided that the benefit is not increased to compensate for the increased price of the environmentally damaging good. The second are people in other countries: if the good, or products made with the good, are exported and the country has sufficient market power that the other countries are unable to switch their source of supply.

Another form of tax shifting is the taxation of goods whose production uses a particularly large amount of an under-taxed factor of production, with the revenue being used to subsidize (or, at least, reduce the taxation of) employment. If environmentally damaging goods make heavy use of under-taxed factors, then their taxation could produce an employment double dividend. To see this, consider a factor that is inelastically supplied (in the sense that the same quantity would be supplied at a lower price) and currently taxed at less than 100%. The situation is clearly not revenue-optimal, as a higher tax on the factor would raise additional revenue without reducing its supply. A direct solution to this situation would be to increase the tax on the income to that factor. However, an indirect partial solution would be to impose some other tax that would result in a fall in the income to that factor, such as a tax on a good whose production made particularly heavy use of that factor. Thus, if capital were inelastically supplied and taxed at less than 100%, and if the production of energy was particularly capital-intensive, a tax on energy could be seen as partly a tax on capital. In this case the imposition of an energy tax that is used to finance a cut in labour taxes shifts the burden of taxation away from labour and towards capital. This would reduce the distortion of labour demand without distorting the supply of capital (which is inelastic), thus creating a strong double dividend.

In applying this analysis, it is important to be sure that the factor really is inelastically supplied. If capital were elastically supplied, perhaps because of the ease of moving it to countries with lower taxes on capital, then the tax shifting would cause a considerable increase in the distortionary cost of the tax system in the form of capital moving abroad. In this case the benefits of the shift in terms of increased employment would be smaller.[14] Bovenberg and De Mooij (1998) have addressed this issue. The desirable level of environmental taxation depends crucially on the elasticity of capital supply and the current rates of capital taxation.

This discussion shows that the shifting of the tax burden away from labour can, in certain circumstances, produce an employment double dividend.

Improving the Efficiency of the Tax System[15]

The previous subsection looked at how a shift in the tax burden could impact on employment. In this subsection we look at whether an employment double dividend can be created without shifting the tax burden.[16] The analysis is interesting because such tax shifting could be difficult if non-labour factors of production are elastically supplied and non-working consumers are protected from bearing the tax burden. The aim is to reduce the tax bur-

---

[14] In fact the distortionary cost of the tax system could be reduced by an environmental tax that fell on labour and was used to finance a reduction in capital taxation!

[15] For a thorough discussion of the principles of taxation and how to increase tax efficiency see ATKINSON A.B. and STIGLITZ J. (1980).

[16] This is not to imply that the effectiveness of the tax system with respect to employment is separate from the shifting of the tax burden. We look at the impacts of shifting and not shifting of the tax burden separately for analytical reasons alone.

den on workers so that labour costs can be reduced and labour demand increased.

If we rule out tax shifting by assuming that all inputs into production are elastically supplied at fixed cost (energy and capital because they are internationally mobile, and labour because the wage is fixed), it can be shown that minimization of production costs requires that all factors are equally taxed. What is actually the case in many countries (particularly Western European countries) is that labour is taxed more heavily than other factors. Thus a shift away from the taxation of labour to the taxation of other factors can be expected to reduce production costs. This will reduce the prices that workers face for the goods they wish to consume. In this case, the move from payroll taxes to taxes on other factors will not be offset by an increase in wages, and employment will increase.

This analysis looks as if it will lead to the existence of an employment double dividend for energy taxation, even without shifting the tax burden. However, the situation is not quite that simple, for two reasons. First, the argument in the previous paragraph was concerned with increasing the tax on all non-labour factors. An increase in energy taxation alone may improve the relative costs of labour and energy, but at the cost of possibly worsening the relative costs of capital and energy. Second, as energy is a produced good (although possibly imported) it may well have already been taxed, and so an additional tax on its use could lead to it being over-taxed. Thus, it is not clear that energy taxation will always lead to an employment double dividend. It is more likely to happen if energy is more substitutable with labour than with capital, as that would make the correction of the relative costs of labour and energy more important than the worsening of the relative costs of energy and capital.[17]

---

[17] KOSKELA E. and SCHÖB R. (2005), have some results of an analysis in which a low enough elasticity of substitution between labour and energy does diminish the positive employment effects substantially. In the same paper the author also shows some conditions under which employment will increase.

Before concluding this discussion of employment creation when there is involuntary unemployment, it is important to note that the possible strong double dividend analyzed here does not apply to all sizes of environmental tax. The arguments presented here have applied to small taxes. As environmental taxes are increased, they increase the distortionary costs of revenue-raising by changing consumer choices – the tax-interaction effect increases – and this effect can outweigh reductions in the distortion of the labour market.

Finally, let us turn to the question of how the wage is determined and whether or not it would in fact be "fixed". This is important because it is possible that the policies discussed here may affect the real wage. The main influence on wages discussed in the literature is trade unions, and we will concentrate on them here. In most models of trade union behaviour, a reduction in unemployment will lead to a higher wage. This will reduce the size of any possible employment double dividend, because any reduction in unemployment will increase the wage, which will in turn increase unemployment. It is, in fact, possible to produce a model in which unemployment cannot be reduced: the entire subsidy to employment is absorbed by an increase in the wage. However, in most of the literature, trades unions are shown to simply reduce the size of any employment double dividend.

The Employment Double Dividend: The Case without Involuntary Unemployment

When the labour market is sufficiently flexible to ensure full employment, the emphasis in the double dividend literature moves away from employment creation and towards the general efficient functioning of markets (*i.e.*, the gross welfare dividend). However, the distortion of the labour market is still a major concern, with the idea that employment taxes tend to reduce the level of labour supply below the optimal level. Hence the use of environmental taxes to partially replace other taxes could increase labour supply and increase measured employment. This section considers the scope for such changes.

The cases of tax shifting discussed above continue to be possible sources of a double dividend, but through a different mechanism. Instead of an increase in labour demand that results from reduced taxation that lowers labour costs, we need to look at supply side incentives in the labour market. In this case, a reduction in labour taxation increases the rewards to working and so increases labour supply, which in a flexible labour market leads to greater employment. The impact of the tax-interaction effect on labour supply is muted by the existence of non-workers in the economy who bear part of the burden of the environmental tax.

Even here, however, the presence of internationally mobile capital can make things go the "wrong" way. Indeed, when capital is able to resist any of the tax burden, the reform can in fact result in lower, not higher real wages. In that case labour supply would be decreased, not increased. This possibility has been explored in two papers by Bovenberg and van der Ploeg (1994*a*, 1994*b*).

In addition, as when there is involuntary unemployment, it is interesting to look at what possibilities there are without tax shifting. Such possibilities again involve improving the efficiency of the tax system. However, the analysis is now different. The arguments presented previously no longer apply as the wage is not fixed at a level that makes supply exceed demand, and so the emphasis is more on increasing labour supply than increasing labour demand.[18] It is necessary to look at the way in which taxes on workers affect their labour force participation. We now turn to an examination of this.

A simple model that is widely used in optimal tax theory is useful for illuminating this question. It is a model in which the only factor of production is labour. In this framework, the only reason for taxing some goods more heavily than others is that their consumption is more closely associated with leisure than oth-

---

[18] Of course, increasing labour demand will increase the wage and so increase labour supply (provided that its supply is not backward bending). However, it turns out that the use of taxes to increase labour demand is an inefficient use of the tax system. Any money is better spent on direct changes to the incentives that workers face.

er goods (in economic terminology, these goods are said to be particularly complementary to leisure). This means that the heavier tax on these goods would implicitly tax leisure and so encourage people to work more, thus reducing the distortion to labour supply produced by the tax system as a whole. So, if an environmentally damaging good was also a good that was consumed in association with leisure, the imposition of a special tax on this good in an economy with otherwise uniform sales taxes could have a double dividend.

It is worth looking at this case in more detail, as it involves a line of reasoning that is quite helpful in understanding the double dividend. One can think of raising the tax on the environmentally damaging good (let us call it energy) and using the revenue to reduce taxes on labour (such as income tax or payroll taxes)[19], as suggested in the double dividend literature. At first sight, this might appear to automatically reduce the distortion of labour supply, but the analysis is more complex. It is not only taxes directly on labour that reduce labour supply, but also taxes on goods that are bought with the income earned by the labour. Thus, increasing the tax on energy also reduces labour supply by reducing the real wage (the tax-interaction effect). However, if energy were particularly complementary to leisure, this effect would be small because people who were deciding whether or not to work more would expect to spend a relatively small proportion of their extra earnings on energy. This means that the disincentive effect of the tax on energy will be less than the incentive effect of reducing taxes on labour, so labour supply would increase and a strong double dividend result.

Note that if the consumption of energy had been associated more closely with labour (*i.e.*, if it had been particularly substitutable for leisure), the result would have been the opposite: the imposition of an energy tax would have reduced labour supply, because energy would have been a relatively larger part of the ex-

---

[19] In an economy without involuntary unemployment, payroll taxes also reduce the incentive to work because they reduce the wage that employers are able to pay their workers.

penditure from the possible extra earnings. In this case, there would not be a strong double dividend. In fact, the environmental tax would have worsened tax distortions in the economy, because of the large negative tax-interaction effect.

A case that has been highlighted in the literature is one that falls between these two possibilities: no goods are particularly associated with either labour or leisure; there is "weak separability" between goods and leisure. In this case, a uniform sales tax is optimal. A very small environmental tax will neither increase nor reduce the distortionary cost of the tax system, but any significant tax will be a move away from the optimum and so increase the distortionary cost. It is this that lies behind the main theoretical result of Bovenberg and Goulder (1996), casting doubt on the existence of a double dividend.

Conclusions from the Theoretical Double Dividend Literature

The following conclusions can be drawn from the review conducted above.

The literature on the double dividend distinguishes between a "weak form" and a "strong form". The strong form, which is the one of interest to policy makers, states that a switch to environmental taxes and away from non-environmental taxes will reduce the welfare cost of raising the current level of government revenue even if their environmental effects are neglected. Hence it is a "gross welfare" dividend in the sense defined earlier. A strong double dividend of this kind cannot occur if the existing tax structure is revenue-optimal. If, however, as is likely in practice, the existing tax structure is not revenue-optimal, a strong double dividend will occur if the new environmental tax moves the tax structure in the direction of revenue-optimality. Therefore, the prospects for a strong double dividend depend on the existing structure of taxation, as well as on other aspects of the economy.

Next we ask when and under what conditions an "employment" double dividend might exist. We need to look separately at two cases: whether or not the labour market is in equilibrium. If it is in disequilibrium, with involuntary unemployment, additional employment is created if the use of environmental taxes to partially replace existing taxes results in an increased demand for labour. If it is in equilibrium, without involuntary unemployment, additional employment is created by increasing labour supply.

There are no necessary or sufficient conditions for environmental taxes to increase employment, but the review has identified factors that make it more likely.

A. The prospects of increased employment *when there is involuntary unemployment* are higher if:

(i) The environmental tax can be passed on to factors that are inelastically supplied and relatively under-taxed.

(ii) Non-working households are large enough in numbers, and are significant as consumers of goods produced with the environmentally intensive inputs that are taxed.

(iii) Through international market power, the environmental tax can raise the price of goods produced with a relatively intensive use of the taxed environmental input. A similar effect would arise if foreign suppliers reduced the price of goods that were subject to environmental taxes when they entered the country.

(iv) Capital is relatively immobile internationally. In this case it can absorb some of the environmental tax and enable the tax to fall less on factors such as labour, enhancing the double dividend effect.

(v) The elasticity of substitution between energy (the environmental input) and labour is greater than the elasticity of substitution between energy and capital.

(vi) The real wage rises little when unemployment falls, so that the reduction in the taxes on labour are not offset by wage rises.

B. *When there is no involuntary unemployment*, conclusions *(i)* to *(iv)* still hold but conclusions *(v)* and *(vi)* are replaced by:

(vii) The environmental tax is levied on goods that are more complementary to leisure than the goods whose taxes are reduced.

These conclusions raise important implications for policy and for the design of empirical models. The empirical models are discussed in section 4, but there are two policy issues that are worth raising here.

First, the importance of capital mobility in determining the existence of an employment double dividend suggests the need for international co-operation in setting environmental taxes. If one country on its own imposes environmental taxes that reduce the return to capital, it could suffer from substantial capital movement. If on the other hand a group of countries imposed such taxes at the same time, there would be less scope for capital to move elsewhere.[20] Against this, however, is the fact that the larger is the group of countries that apply the taxes, the smaller is the remaining set of countries that will have to pay the shifted taxes and so the smaller the amount of tax that can be shifted.[21]

Second, the literature does not specifically deal with the practical question of which taxes on labour should be reduced to get the largest employment double dividend. Should it be income taxes or social security taxes? Intuitively, it seems likely that it should be social security taxes because they are more closely linked to employment than income taxes, which can cover non-labour incomes and are progressive (thus bearing less heavily on the incomes of lower-paid workers). This intuition has been tested and results reported in section 2.3.

Third, and related to the above point, we need to consider the wisdom of Green Tax reform when the dividends are generated as a result of shifting the tax burden to non-workers. If most of these are pensioners or unemployed persons, there is a negative distributional impact from the tax reform, which many would consider undesirable.

Finally, we can also ask which of the conditions listed above will promote *both* the gross welfare and employment double div-

---

[20] To some extent, this argument applies also in considering international market power, both in terms of being able to increase the price of exports and in terms of being able to reduce the price of imports. This market power will be greater for a group of countries acting together than for a single country. It is worth mentioning that these concerns go beyond environmental policy. They are general issues related to domestic capital taxation.

[21] International co-operation may also be useful in minimizing the loss of international competitiveness that could result from introducing environmental taxes. International competitiveness is not considered in the theoretical literature, which assumes that exchange rates adjust to maintain equilibrium in the balance of payments. However, it is captured in the empirical models – see the next section.

idends. From the previous discussion and other literature one can say that factors *(i)*, *(iv)*, *(vi)* and *(vii)* are likely to also result in a gross welfare dividend, although this is not guaranteed. In general it is much more difficult to ensure a gross welfare dividend than an employment dividend.

3. - An Empirical Analysis of Climate Related Green Taxes

This section reports on a three aspects of climate related green taxes. First it analyzes the relative impacts of carbon taxes *versus* permits as instruments for climate regulation. Second it considers the implications of taxes and permits on the adoption of new technologies and third it reviews the literature on the environment and employment impacts of carbon taxes in the European context, investigating the possibility of a double dividend through a carbon tax.

3.1 *Carbon Taxes* versus *Permits*

The debate on price *versus* quantity instruments for tackling climate change has continued for decades. Although the introduction of a European emission trading scheme within the Kyoto framework could have represented a pragmatic resolution to this discussion, the need to engage the US and other large-emitters fast-growing countries, has led some economists to suggest a global carbon tax as a possible way around the impasse (see, for example, Stiglitz, 2006). One of the key issues here is the impact of uncertainty on the choice of instrument. As we noted earlier, the problem was stated analytically by Weitzman (1974); subsequently, numerical developments have been carried out in Pizer (1999), and Newell and Pizer (2003). These questions are discussed further in the paper in the present issue by Galarraga and Ansuategi (forthcoming). Here we consider one special dimension of the problem, namely uncertainty in the context of framing a climate policy.

In a recent paper Bosetti *et* al. (2007) study how uncertain abatement costs and uncertain climate sensitivity (which ultimately reflects on climate damages), jointly although independently, affect optimal choices when a stabilization target is imposed through a price instrument and compare it to the case when a quantity instrument is adopted. While the presence of uncertain abatement costs pushes risk adverse individuals to prefer the price instrument, the randomness of climate damages introduces an opposite bias towards the quantity instrument. The paper analyses how these two competing forces combine and comments on the resulting optimal policy choices for a risk adverse individual.

For the purpose of the analysis the authors apply Monte Carlo simulations to WITCH (World Induced Technical Change Hybrid), an optimal growth integrated assessment model with a fairly detailed energy sector. First, the policy scenarios under examination – Cap & Trade and Carbon Tax – are analysed in a deterministic setting so as to verify consistency with the Weitzman result (1974) – *i.e.* that in the absence of uncertainty, price- and quantity-based market instruments are equivalent in their economic and environmental impacts. Subsequently, the analysis is carried out for uncertain abatement costs and climate sensitivity. With a Monte Carlo simulation it evaluates the effect of uncertainty on endogenous GDP, consumption, CO_2 emissions, R&D investments in the energy sector and investment in clean electricity generation technologies, under price and quantity policy instruments.

Results show that uncertainty leads to GDP and consumption with higher means and lower variances under the price instrument than under the quantity instrument. In this sense the price instrument stochastically dominates the quantity instrument with respect to GDP and consumption. It is particularly interesting that this result is not reversed when uncertainty on climate damages is introduced into the model. Emissions on the other hand are constant under the Cap & Trade scenario, while they adjust to random differences in abatement costs under the Carbon Tax scenario, not necessarily satisfying the limits sought by a stabilization target. The explanation for this result lies in the fact that the effect of uncertain abatement costs is entirely reflected in economic growth in the Cap & Trade case, while under the Tax scenario, the carbon tax offers a safety valve in case of higher than expected abatement costs.

Thus while the tax option dominates with respect to GDP and consumption, it does less well with respect to achieving emissions reduction targets. The costs of not achieving the emissions targets, however, do not turn out to be that high: even in case of higher than expected climate damages, the penalty for non compliance to the environmental target is relatively small when carbon taxes are very high as in the stringent stabilization scenario considered here. This stems from the fact that one of the results of greatly reducing carbon emissions is precisely that of hedging against worse than expected climate change consequences by keeping carbon concentrations under control. Inter-temporal discounting further reduces the cost of slightly missing the environmental target. These issues, together, make the penalty rather small. Energy R&D investments appear to be higher under the Tax scenario (when productivity of R&D is higher than average, higher than average investments are induced by the carbon tax) but to display higher variance under the Cap & Trade scenario (notwithstanding the effectiveness of R&D, the target has to be achieved and at least some investments have to be undertaken).

Finally, investments in renewables for electricity generation shows a higher mean and variance under the quantity instrument. Although the difference with the price instrument is not large it is an issue of concern and we consider it below.

3.2 *Carbon Taxes and Investment in Low Carbon Technology*

Many climate policy experts have stressed the importance of R&D in new technologies as a solution to the climate problem, much more than actions that increase the efficiency of existing technologies. Yet taxes or permits are clearly necessary to provide the underlying incentive for more investment in R&D. As was found in the paper by Bosetti *et* al. (2006) taxes appear to provide slightly less investment in renewables through R&D, although in that study the impact was small relative to permits.

In a paper looking at the same issue from a theoretical perspective Golub *et* al. (2008) explore the implications of risk-averse firm hypothesis using a mean-variance analytical framework. The primary result is that price instruments, including emissions taxes and cost caps, may reduce willingness to deploy new technologies. This conclusion is coherent with results presented in the previous paper. Moreover, the paper finds that, in order to ensure the same innovation rate (or the same level of pollution), emissions taxes should be relatively higher than the expected (mean) value of the (uncertain) price of emissions allowances. In other words, taking into account risk-averse behavior, price based policy should be more stringent in order to ensure same environmental results as a less stringent cap. Taking into account behavioral responses the regulator should raise taxes relatively higher than expected price in a carbon market based on permits.

These findings suggest that perhaps neither carbon taxes nor permits by themselves will be enough to provide the incentives needed to bring about the required investment in developing new technologies consistent with achieving the desired climate stabilization targets. Other instruments that are likely to be necessary include public funds for basic research (which has significant spillover externalities) as well as protection of intellectual property rights for investment in low carbon technology.

3.3 *Carbon Taxes and the Double Dividend*

This section reports the results of estimating the employment double dividend for the European Union in relation to the introduction of a carbon tax. Several models have been used in this work and their assumption results are summarized in Table 2 below. For more details see Heady *et* al. (2000).

Key Aspects of the Empirical Models

The discussion of the theoretical literature in section II suggests that the following features of the economy are important in assessing the likelihood of a double dividend. It is worthwhile to look at the extent to which the models capture each of these features.[22]

Existing tax structure. This is captured in detail by each model.

Complementarity of consumption goods to leisure. This has not been captured in any of the models as they all assume that all goods are equally complementary.

The pattern of factor intensities of production for different goods. This has been captured in detail by each of the models.

The characteristics of non-worker consumers. These are not well captured in any of the models as they all appear to use the representative household approach. The only non-worker consumer is the rest of the world, and all models assume that the EU countries have some monopoly power in trade. This gives the models some ability to pass on energy tax increases to foreigners and thereby create a larger double dividend in the EU.

International mobility of factors of production. This is really only an issue for capital. None of the models explicitly addresses the issue of the international mobility of capital.[23] If capital is mobile in this way, it will seek the highest return, and investment in anyone country must respond to differences between domestic and international rates of return. An increase in energy prices, which can be passed on to capital and thereby reduce the rate of return on capital, should imply a reduction in domestic investment. This reduction in turn will raise domestic rates of return until the international and domestic rates are equalized. Hence capital will

---

[22] PATUELLI R. *et* AL. (2005) have carried out a similar but wider exercise recently and have concluded that the following factors are critical in determining difference in the results of empirical models: tax type, recycling policy and whether the model is derived from micro foundations or is more of a macro type.

[23] None of the CGE models has examined this but some analytical GE models have. See, for example, BOVENBERG A.L. - GOULDER L.H. (1997).

not bear part of energy tax, and the tax shift will not result in as big a gain in employment as when capital is immobile. In view of this, we believe that *all* the models could exaggerate the impacts of a shift in taxation from labour to energy in terms of increased employment. If capital is indeed mobile the burden of increased energy taxation could not be passed on to capital and would be borne by energy and labour, reducing the size of the double dividend.

The responsiveness of labour demand or supply to changes in labour taxes. The modeling of the labour market is divided into those models that assume full employment or voluntary unemployment (GEM-E3 and HONKATUKIA) and those that assume involuntary unemployment (HERMES and EUROGEM). The former may generate a "double dividend" in the sense that employment increases as the incentives to supply labour become stronger. However, as the people who have moved into employment were previously voluntarily unemployed, the benefit to society is very different from the benefit created when involuntary unemployment is reduced. The two sets of employment effects are therefore not really comparable, although they are frequently compared.

The Elasticities of Substitution Between Labour, Capital and Energy. We noted that the greater the elasticity between labour and energy, relative to the elasticity between capital and energy, the more likely it is that an employment double dividend will exist. In general the models have Allen elasticities that reflect this and therefore make the possibility of a double dividend quite strong.[24]

Analysis of the Impacts of the 1992 EU Energy Tax Proposal

There are now many models that have looked at the impacts

---

[24] If the cost function for the firm is $E = G(X_1, X_2, ..., X_N, Y)$, where the Xs are inputs and Y is output, the Allen elasticity of substitution between inputs i and j is given by $\Phi_{ij} = [G \cdot G_{ij}/G_i G_j]$. The cross price elasticity between inputs i and j is given by $Eij = \Phi_{ij}M_j$, where M_j is the share of input j in total cost.

TABLE 2

MAIN FEATURES OF SIMULATION MODELS USED
IN EUROPEAN EMPLOYMENT/CARBON TAX STUDIES

Model	Key Economic Assumptions	Special Points
HERMES	CGE model with unemployment. Uses nested CES production functions. National and EU applications.	Detailed development at national level in EU. Structure is transparent. Real wages determined by productivity growth and unemployment.
EUROGEM	Similar to HERMES. National and EU applications.	EU-wide model. Real wages now also depend on trade union bargaining objectives, which are a function of employment and real income differentials between workers and the unemployed.
GEM-E3	Classical CGE model with full employment. Run at EU12 and EU15 level	Structure has more emphasis on consistency with general equilibrium theory than with detailed estimation of structural equations. Information on model structure is cursory.
E3ME	Econometric model with less basis in economic theory. Assumes unemployment.	No production functions specified; only input demand functions with increasing returns. Cannot derive underlying productions functions from them.
HOKATUKIA	Model for Finland only. Dynamic CGE model with relatively simple structure and full employment.	Firms are imperfectly competitive, which allows some of the tax to be passed on in higher prices. Implications of environmental tax for overall efficiency of economy remain unclear.
LEAN-TCM	Similar structure to HERMES with unemployment.	Real wage depends on tightness of labour market.

Notes: the EU12 are Belgium, Denmark, France, Germany, Greece, Ireland, Italy, Luxemburg, Netherlands, Portugal, Spain and the UK; the EU15 are the EU12 *plus* Austria, Sweden and Finland.

of green tax reforms, but is difficult to come across a range that have addressed the same reforms. One exception to this is the 1992 EU energy tax proposal. For this reason we look in some detail at the results of these models.[25] In this, a 50:50% mix of carbon and energy taxes is applied at the level of $3/barrel of oil equivalent (b.o.e.) in the first year and rises to $10/b.o.e. in sev-

---

[25] A similar comparative assessment of the 1997 tax reforms was conducted by JANSEN H. and KLAASSEN G. (2000).

en years.[26] This is a revenue-neutral change, with tax revenue being recycled through reduced employers' social security contributions. The models that have been run for this option are E3ME, GEM-E3, LEAN-TCM and EUROGEM. The E3ME is not run for exactly the same scenario, as it increases taxes from $1 per barrel of oil equivalent to $13 in 11 years. The LEAN-TCM also has slightly different tax increases than the others. The E3ME model is only run for the UK.

In spite of these limitations, a comparison of the results is instructive. Table 3 presents the main findings. The following points are worth noting:

A. The models all predict GDP increases, but they differ considerably in terms of the size of the increase, with LEAN-TCM producing the biggest increase in the final year, followed by EUROGEM, GEM-E3 and E3ME. The use of GDP is not, of course, a perfectly reliable indicator of a gross welfare dividend, which is better measured in terms of "overall consumption" or non-environmental welfare (as measured by an equivalent variation).

B. The time profile of the increases also varies. EUROGEM picks up much faster than E3ME. LEAN-TCM does not appear to have increased impacts over time at all. We do not have data on the time profile for GEM-E3.

C. The employment increase is greatest for LEAN-TCM, followed by E3ME. We attribute the high value in LEAN-TCM to the low wage elasticity with respect to unemployment. This means that when taxes are lifted and employment demand increases, the real wage does not increase by much to negate the tax advantage. The E3ME effect is probably due to the increasing returns to scale assumption cited earlier, and partly to a greater substitutability of labour for energy. It is noteworthy that, in terms of employment, EUROGEM produces similar results to E3ME for the EU12. GEM-E3 has a much smaller employment impact.

D. The employment/GDP ratios vary a great deal. E3ME has

---

[26] A tax of $10 per b.o.e. amounts to a tax of around $4 per ton of CO_2. This is based on *(a)* one b.o.e. is equal to 5.5 gigajoules, *(b)* one gigajoule of petroleum products generate 72 kg of CO_2.

much the highest ratio, followed by EUROGEM and GEM-E3. This suggests that the substitution potential in the three models differs quite a lot, with E3ME having the greatest and GEM-E3 the lowest.

    E. E3ME generates a fall in prices, whereas the other two show a small increase in the price level, indicating that a shifting of the tax to the rest of the world is unlikely to be big.

TABLE 3

IMPACTS OF AN ENERGY/CARBON TAX IN EU12:
SOME COMPARATIVE RESULTS (FIGURES ARE
PERCENTAGES OVER BASELINE)

Year	Model	Countries	GDP increase	Employ-ment increase	Carbon decrease	Energy decrease	Price increase	Employ-ment/ GDP
YR 1	E3M3	UK only	0.02	0.12	0.33	N/A	0.00	0.10
	GEM-E3	UK only	N/A	N/A	N/A	N/A	N/A	N/A
	GEM-E3	EU12	N/A	N/A	N/A	N/A	N/A	N/A
	LEAN/TCM	EU12	0.47-1.4	0.7-2.24	4.1-4.8	N/A	N/A	0.2-0.8
	EUROGEM	EU9	0.00	0.20	5.00	N/A	N/A	N/A
YR 3	E3M3	UK only	0.05	0.47	1.31	N/A	-0.03	0.42
	GEM-E3	UK only	N/A	N/A	N/A	N/A	N/A	N/A
	GEM-E3	EU12	N/A	N/A	N/A	N/A	N/A	N/A
	LEAN/TCM	EU12	N/A	N/A	N/A	N/A	N/A	N/A
	EUROGEM	EU9	0.60	1.08	6.92	N/A	N/A	0.48
YR 10	E3M3	UK only	0.12	2.59	4.51	N/A	-0.22	2.47
	GEM-E3	UK only	0.30	0.53	N/A	-7.72	2.25	0.23
	GEM-E3	EU12	0.15	0.37	10.34	-5.08	3.89	0.22
	LEAN/TCM	EU12	0.4-2.1	0.8-3.2	6.2-7.6	N/A	N/A	0.4-1.1
	EUROGEM	EU9	0.90	2.20	16.00	N/A	N/A	1.30

Sources: CAMBRIDGE ECONOMETRICS (1998); CAPROS P. *et* AL. (1996); BAYAR A. (1998); WELSCH H. (1996).
Notes: E3M3 analyses a carbon tax starting at $1 per b.o.e., rising to $13 in year 11; For E3ME and EUROGEM the last row is for year 11; GEM-E3 and EUROGEM analyse a carbon tax starting at $3 per b.o.e. and rising to $10 in year 7; The year 3 value for EUROGEM is interpolated; GEM-E3 EU figures are estimated from individual country data, using appropriate weights; The EU9 are the EU12 without Germany, Greece and Luxemburg.

    To sum up, the empirical analysis has shown that the models differ in a number of ways that the theoretical analysis suggested would influence the likelihood of employment creation. It is therefore interesting to note that *all* the models suggest that the partial replacement of taxes on labour by taxes on energy increases employment and reduces carbon emissions. However, there is considerable variation between the models in the size of these effects.

It is impossible to use theoretical analysis to determine which of the many differences between the models are responsible for the differences in predicted employment creation. Instead, we must turn to a sensitivity analysis, which reports on numerical experiments to investigate which factors are most important in determining the amount of employment created.

Sensitivity Analysis

Capros *et* al. (1996) provide some idea of the sensitivity of the GEM-E3 model to some parameter variations. A selection of their results is shown in the Table 4. Unfortunately, they do not provide details of the actual parameter values they use. Nonetheless, the table gives some hints as to what might be important. The first line of the table gives the base case results. The second line reports the consequences of making the wage rate strongly dependent on the level of employment, which reverses the GDP change and eliminates the employment gain: because of the wage rise, the energy taxes distort production with no employment gain. The third and fourth lines show the implications of altering the elasticity of substitution between labour and materials, showing that a high elasticity promotes employment growth and reduces energy use. This is what one would expect from labour subsidies.

The fifth line shows the effect of regarding the export market as more competitive, producing a smaller employment gain despite the continued reduction in energy use. It also shows that the GDP growth is substantially reduced. This illustrates the importance of tax shifting, in this case to the rest of the world, in producing a double dividend.

Finally, the last two lines show the effect of altering the substitutability between the capital/electricity aggregate and the labour/materials aggregate. The effects on employment are relatively modest, but interestingly show that a higher elasticity of substitution produces a smaller employment effect. This is presumably because the higher elasticity creates a greater distortionary effect between capital and (non-electrical) energy, thus reducing efficiency, as witnessed by the much smaller gain in GDP.

TABLE 4

SENSITIVITY OF GEM-E3 MODEL PREDICTIONS TO KEY
PARAMETERS (FOR FRANCE) SIMULATION OF EUROPEAN
COMMISSION'S 1992 CARBON/ENERGY TAX PROPOSAL

	Change in GDP	Change in Energy	Change in Employment
Base case	0.16%	-4.40%	+91,000
Inelastic Labour	-0.14%	-4.25%	0
Low labour-materials substitution	0.07%	-1.21%	+33,000
High labour-materials substitution	0.18%	-6.42%	+116,000
Competitive exports	0.05%	-4.70%	+68,000
Low capital-labour substitution	0.16%	-4.40%	+90,000
High capital-labour substitution	0.05%	-4.70%	+68,000

In addition, Denise Van Regemorter (Department of Economics, University of Leuven, private communication) has informed us that GEM-E3 does not produce a double dividend for any country if the assumption of EU monopoly power in international trade is dropped, because of the EU's loss of competitiveness. In other words, in this model, the double dividend only exists if some of the tax can be shifted to overseas consumers. This emphasizes the importance of looking carefully at the modelling of non-worker consumers, something that is frequently overlooked in the literature. However, it should be noted that EU monopoly power is not required to produce a double dividend in all the models. Terry Barker (1994) (Cambridge University; private communication) has confirmed that the E3ME double dividend does not depend on being able to pass on costs in export prices.

An additional source of evidence on sensitivity was provided by Ali H. Bayar (Université Libre De Brussels), who carried out some simulations for us with EUROGEM. The results are shown in Table 5. The first row of Table 5 reports the percentage changes to employment and carbon dioxide emissions in the 20th year after the policy change, based on the standard model assumptions with revenue recycled through reductions in social security taxes. The second and third rows show the changes if the elasticity of substitution between labour and the capital-energy aggregate is changed. As expected from the theory, increasing this elasticity substantially increases the employment benefit, as labour substitutes more easily for energy in response to the tax changes. Halving the elasticity actually eliminates the employment double dividend completely.

The third and fourth rows of Table 5 also illustrate the theoretical results, by showing the importance of the elasticity of substitution between capital and energy. As expected, reducing the elasticity increases the employment double dividend, as the energy tax has a smaller distortionary effect on the quantity of capital used in production. Doubling the elasticity removes the double dividend.

The last row of Table 5 reports on the effect of recycling the revenue through reductions in labour income tax instead of social security taxes. This confirms the intuition of the theoretical section: social security taxes are the best form of revenue recycling for obtaining an employment double dividend.

TABLE 5

SENSITIVITY OF EUROGEM MODEL PREDICTIONS
TO MODEL ASSUMPTIONS SIMULATING THE IMPOSITION
OF A US $10/TONNE CARBON/ENERGY TAX

	Change in Employment after 20 years	Change in Carbon Dioxide after 20 years
Base Case	+ 0.61%	-17.93%
Doubling of top level elasticity: KE *vs.* L	+3.41%	-18.84%
Halving of top level elasticity: KE *vs.* L	-0.08%	-17.24%
Doubling of second level elasticity: K *vs.* E	-1.14%	-26.78%
Halving of second level elasticity: K *vs.* E	+1.25%	-12.26%
Recycling revenue through labour income tax	-2.12%	-18.55%

Unfortunately, we have been unable to obtain any sensitivity results for the effect of international capital mobility, which the theoretical analysis suggests could be of considerable importance. As reported earlier, the models do not represent international capital mobility directly, although some do make investment depend on the rate of return on capital. However, none of the modellers have reported the sensitivity of their results to changes in the parameters that link investment to the rate of return.

An Assessment of the Danish Carbon Tax

The Danish government introduced a carbon tax in 1996, with rates of 5DKK/ton CO_2 for heavy process, 50 DKK for light processes and 100 DKK for space heating. The rates for heavy and light processes were increased by 2000 to 25DKK and 50DKK respectively by 2000. Exchange rates have varied over this period, but the 2000 rates amount to a tax range of between \$4-14/ton CO_2.[27] There are, however, considerable exemptions to the taxes, with energy intensive industries in heavy processes paying only around 3DKK/ton CO_2.

The carbon tax was recycled through four channels: employers contributions to social security (reductions in the payroll tax); employers contributions to pensions (reductions in the ATP); subsidies for investment in new energy efficient technology; and a special fund for small enterprises.

The Danish government reviewed the experience of the tax. As only a presentation of the work was available to the author, a full comparison with the other studies is not possible. Nevertheless, the results are of interest and complement those from the studies discussed above.

The following are the main findings from the Danish experience:

A. The estimated CO_2 reductions from the carbon tax in 2005

---

[27] Details were provided by Mr. Larsen from the Danish Ministry of Taxation and are available on: *http://www.nmr.ee/dokumendid/nordic_forum/larsen.pdf*

were estimated at around 2 percent, which is a relatively small contribution. This is probably the result of significant tax reductions to energy intensive industries (50 percent of emissions are caused by energy intensive industries that pay only 20 percent of energy taxes). At the same time, however, another 1.8 percent reduction in CO_2 emissions was attributed to the subsidies mentioned above.

B. The impacts on employment were not reported, but are estimated to be positive and small (Mr. Larsen, private communication).

C. The tax differentiations outlined above were considered necessary to maintain international competitiveness in the energy intensive sectors. (Again the details of the analysis of this are not available).

D. The administrative costs of the tax to companies amount to 1-2 percent of the revenue, but the costs of applying for the subsides are much higher: around 3-9 percent of the amount of the subsidies.

Conclusions on the Empirical Evidence on the Employment Double Dividend

Almost without exception these European models find that a switch in taxation from labour to carbon/energy will increase employment and reduce carbon emissions. At the same time it will increase GDP. Hence there is some agreement on this "good news". The differences are about the size of the impacts. For the 1992 proposed carbon/energy tax, which rises to $10 per barrel of oil equivalent over about 7 years, the size of the employment impact ranges from 0.4 to 2.6 percent by the end of that period. This is for various groupings of EU countries and should therefore be treated with caution, but it is still instructive about the range of estimates. With around 140 million people employed in the EU12, it translates into a range of from half a million extra jobs to over 3.5 million. The range of impacts on carbon reductions is also huge, from 5 to 16 percent. The GDP increases range from 0.4 to 2.2 percent. It is also interesting that more recent work, such as that carried out by national governments *ex post*, reveal impacts at the lower end of these ranges. The Danish study presented above is an example.

As expected, and as emphasized in the theoretical literature, the degree of substitution between inputs is important in determining the additional employment created. The larger the elasticity of substitution between labour and energy, the larger was the employment increase. Also, the smaller the substitution elasticity between capital and other inputs, the larger was the employment increase. These reflect the efficiency effects of reducing the high tax on labour. The importance of efficiency was also demonstrated in EUROGEM by the fact that the use of energy tax revenue to reduce income taxes, rather than payroll taxes, reversed the employment gain. This reflects the fact that income taxes are not so closely related to employment as are payroll taxes.

The role of shifting the tax burden was also demonstrated in GEM-E3 by two results: employment gains were reduced with increased export competition, and employment gains disappeared completely if none of the tax can be shifted abroad. Unfortunately, none of the models captured the possibility of shifting taxes between different groups within society, such as those on pensions. More importantly, none of the models was able to indicate the sensitivity of the results to changes in the ability to shift the tax burden onto capital. The theoretical analysis suggests that this is very important, and the investigation of this with numerical models should be a high priority for further research.

We should also not forget that there is a tension between the employment and environmental benefits on the one hand and the distributional impacts on the other. As we showed, the factors that make the Green Tax more effective in the former areas are also the factors that could have disproportionate negative effects on the groups like pensioners and unemployed.

For all the reasons given above we would urge caution in assuming that the actual impact of green taxes would have a positive employment effect. We can only weakly suggest that countries could benefit from a reduction in the employment tax and switch to a carbon tax. The theoretical analysis shows that a simple double dividend view is *naïve*; the reality is much more complex, and many of the assumptions that have to hold for the employment double dividend to occur are difficult to justify. In the end the employment double dividend turns out to be an empirical issue. The empirical work is indicative of a small double dividend but, painstaking as it is, a number of the key linkages are left out. Hence policy makers would be justified in treading carefully in this area. An initial move to switch taxation may be tried experimentally and, if successful, extended.

Finally we note that the models reviewed above pertain exclusively to Europe. Results from the US and some developing countries are less supportive of an employment double dividend (see e.g. Bovenberg and Goulder, 1996). This suggests that the prospects for a double dividend might be better in Europe than in the US. Perhaps this reflects greater inefficiencies in labour markets in Europe, where labour appears to be "overtaxed" relative to capital in the sense that the marginal excess burden per dollar of revenue is higher for an incremental increase in the labour tax than for an incremental increase in the capital tax. In the US, the situation is the opposite. This means that shifting the tax burden from labour to capital works in favour of not only the employment dividend but also the gross welfare dividend in Europe, while it works against the gross welfare dividend in the US Conversely, policies that promote the gross welfare double dividend in the US – lower capital taxes – work against the employment double dividend.

4. - Conclusions

This paper has surveyed the literature on environmental taxation in the context of the climate change problem. The history of environmental taxation goes back about 75 years and has a good intellectual pedigree. Predictably its practical implementation has been more patchy and messy than the theory would suggest. Nevertheless there is a strong movement to use environmental taxes as a means of regulation and climate is a major area where the tax has both been introduced (to a limited extent) and where it is being proposed. The paper reviews the increased use of taxes as instruments of environmental management and notes the strengths and weaknesses of the applications.

The paper has shown that the performance of taxes relative to permits depends on the presence of uncertainty. Given that there is uncertainty with respect to both costs of mitigation and the damages from climate change, one would expect taxes to have different results from permits and we find that this is indeed the case. Taxes have a higher GDP mean and a lower variance than permits when both kinds of uncertainty are taken into account. On the other hand, taxes result in lower investment in renewables. The simulation results suggest the difference between taxes and permits are small but this needs to be verified. In general research suggests that taxes may need to be higher than the mean value of permit prices if they are to be as effective as the latter in terms of investment in new technology. We would also conclude that the technology objective needs other instruments, in addition to taxes or permits.

The paper also reports on simulation studies of the possible double dividend implications of a carbon tax combined with a reduction in other taxes. We should recognize that these are *simulation* studies – *i.e.* they are not based on actual observed data and to that extent they have their limitations. Nevertheless they are indicative of possible trends and it is important to note that, almost without exception, the models for the EU find that a switch in taxation from labour to carbon/energy would increase employment and reduce carbon emissions. At the same time they would increase GDP. Hence there is some agreement on this "good news". The differences are about the size of the impacts. With around 140 million people employed in the EU12, the effects translate into a range of from half a million extra jobs to over 3.5 million. The range of impacts on carbon reductions is also huge, from 5 to 16 percent. The GDP increases range from 0.4 to 2.2 percent.

BIBLIOGRAPHY

ATKINSON A.B. - STIGLITZ J., *Lectures on Public Economics*, New York, McGraw-Hill, 1980.

BARKER T., «Taxing Pollution Instead of Employment: Greenhouse Gas Abatement through Fiscal Policy in the UK», Department of Economics, *Discussion Paper*, no. 94-10, Birmingham (UK), University of Birmingham, 1994.

BAYAR A., *Can Europe Reduce Unemployment through Environmental Taxes?*, paper presented to the Twelfth International Conference on Input-Output Techniques - International Input-Output Association, New York, 1998.

BOSETTI V. - CARRARO C. - GALEOTTI M. - MASSETTI E. - TAVONI M., «WITCH: A World Induced Technical Change Hybrid Model», *The Energy Journal*, no. 27(2), special issue on Hybrid Modeling of Energy-Environment Policies: Reconciling Bottom-up and Top-down, 2006, pages 13-38.

BOSETTI V. - GOLUB A. - MARKANDYA A. - MASSETTI E. - TAVONI M., *Abatement Cost Uncertainty and Policy Instrument Selection Under a Stringent Climate Policy*, A Dynamic Analysis, Milano (IT), FEEM, no. 15.2008, 2007.

BOVENBERG A.L. - DE MOOIJ R., «Environmental Taxes, International Capital Mobility and Inefficient Tax Systems: Tax Burden versus Tax Shifting», *International Tax and Public Finance*, no. 5, 1998, pages 7-39.

BOVENBERG A.L. - GOULDER L.H., «Optimal Environmental Taxation in the Presence of Other Taxes: General Equilibrium Analyses», *American Economic Review*, no. 86, 1996, pages 985-1000.

— - —, — - —, «Costs of Environmentally-Motivated Taxes in the Presence of Other Taxes: General Equilibrium Analyses», *National Tax Journal*, no. 50, 1997, pages 59-88.

BOVENBERG A.L. - VAN DER PLOEG F., «Green Policies and Public Finance in a Small Open Economy», *Scandinavian Journal of Economics*, no. 96(3), 1994*a*, pages 343-363.

— - —, «Environmental Policy, Public Finance and the Labor Market in a Second-Best World», *Journal of Public Economics*, no. 55, 1994*b*, pages 349-390.

— - —, «Optimal Taxation, Public Goods and Environmental Policy with Involuntary Unemployment», *Journal of Public Economics*, no. 62(1-2), 1996, pages 59-83

CAMBRIDGE ECONOMETRICS, *Industrial Benefits from Environmental Tax Reform in the UK*, Cambridge, Econometrics Technical Report, no. 1, Sustainable Economy Unit of Forum for the Future, 1998.

CAPROS P. - GEORGAKOPOULOS P. - ZOGRAFAKIS S. - PROOST S. - VAN REGEMORTER D. - CONRAD K. - SCHMIDT T. - SMEERS Y. - MICHIELS E., «Double Dividend Analysis: First Results of a General Equilibrium Model (GEM-E3) Linking the EU-12 Countries», in CARRARO C. - SINISCALO D. (eds.), *Environmental Fiscal Reform and Unemployment*, Dordecht, Kluwer Accademics, 1996.

ECOTEC, «Study on the Economic and Environmental Implications on the Use of Environmental Taxes and Charges in the European Union and its Member States», *http://europa.eu.int/comm/environment/enveco/database_env_taxation.htm*, 2001.

GALARRAGA I. - ANSUATEGI A., «Carbon Pricing as an Effective Instrument of Climate Policy: Searching for an Optimal Policy Instrument», *Rivista di Politica Economica*, forthcoming.

GOLUB A. - DUDEK D. - STRUKOVA E., «Risk-Averse Firm and New Technologies», in GOLUB A. - MARKANDYA A. (eds.), *Modeling Environment-Improving Technological Innovations under Uncertainty*, London, Routledge, 2008.

GOULDER L.H., «Environmental Taxation and the Double Dividend: A Reader's Guide», *International Tax and Public Finance*, no. 2, 1995, pages 157-183.

HEADY C. - MARKANDYA A. - BLYTH W. - COLLINGWOOD J. - TAYLOR P.G., *Study on the Relationship between Environmental/Energy Taxation and Employment Creation*, Final Report, Prepared For The European Commission: Directorate General Xi Contract: B4-3040/98/00016/MAR/B1, 2000.

JANSEN H. - KLAASSEN G., «Economic Impacts of the 1997 EU Energy Tax: Simulations with Three EU-Wide Models», *Environmental & Resource Economics*, no. 15(2), 2000, pages 179-197.

KOSKELA E. - SCHÖB R.,. «Optimal Capital Taxation in Economies with Unionized and Competitive Labour Markets», *Oxford Economic Papers*, vol. 57(4), 2005, pages 717-731.

LIGTHART J.E. - VAN DER PLOEG F., «Environmental Policy, Tax Incidence, and the Cost of Public Funds», *Environmental and Resource Economics*, no. 13(2), 1999, pages 187-207.

MARKANDYA A., «Policy Failures as a Cause of Environmental Degradation», in MALER K.G. - VINCENT J. (eds.), *The Handbook of Environmental Economics*, General Editors, Arrow K. - Intriligator M., Amsterdam, North Holland/Elsevier Science, 2005.

NEWELL R.G. - PIZER W.A., «Regulating Stock Externalities Under Uncertainty», *Journal of Environmental Economics and Management*, no. 45, 2003, pages 416-432.

OECD, «Environmental Taxes in OECD Countries: A Survey», OECD, *Monographs*, no. 71, Paris, OECD, 1993.

PARRY I.W.H., «A Second Best Analysis of Environmental Subsidies», *International Tax and Public Finance*, no. 5, 1998, pages 157-174.

PATUELLI R. - NIJKAMP P. - PELS E., «Environmental Tax Reform and the Double Dividend: A Meta-Analytical Performance Assessment», *Ecological Economics*, no. 55(4), 2005, pages 564-583.

PEARCE D., «The Role of Carbon Taxes in Adjusting to Global Warming», *Economic Journal*, no. 101, 1991, pages 938-948.

PIGOU A., *The Economics of Welfare*, London, MacMillan, 1932.

PIZER W., «Optimal Choice of Policy Instrument and Stringency under Uncertainty: The Case of Climate Change», *Resource and Energy Economics*, no. 21, 1999, pages 255-287.

REPETTO R. - DOWER R.C. - JENKINS R. - GEOGHEGAN J., *Green Fees: How a Tax Shift Can Work for the Environment and the Economy*, Washington DC, World Resources Institute, 1992.

SANDMO A., «Optimal Taxation in the Presence of Externalities», *Swedish Journal of Economics*, no. 77(1), 1975, pages 86-98.

SNOW A. - WARREN JR. R.S., «The Marginal Welfare Cost of Public Funds: Theory and Estimates», *Journal of Public Economics*, no. 61, 1996, pages 289-305.

STIGLITZ J., «A New Agenda for Global Warming», *The Economists' Voice*, no. 3(7), 2006.

TIETENBERG T.H., «Economic Instruments for Environmental Regulation», *Oxford Review of Economic Policy*, 6(1), 1990, pages 17-33.

VON MOLTKE A. - MCKEE C. - MORGAN T., *Energy Subsidies: Lessons Learnt in Assessing Their Impact and Designing Policy Reforms*, Sheffield (UK), Greenleaf Publishing, 2004.

WELSCH H., «Recycling of Carbon/Energy Taxes and the Labour Market», *Environmental and Resource Economics*, no. 8, 1996, pages 141-155.

WEITZMAN M.L., «Prices vs. Quantities», *Review of Economic Studies*, no. 41, 1974.

Market-Based Instruments in CEE Countries: Much Ado about Nothing

Milan Ščasný - Vojtěch Máca*

Charles University Environment Center

This paper overviews the use of environmentally related levies and other economic instruments in air quality and climate change policies in Central and Eastern European EU countries. It is shown that their overall effectiveness and efficiency is quite low and they help to secure revenues for dedicated environmental funds only. We pay a special attention to the implementation of 2003/96/EC Directive in the Czech Republic that presents the unique example of a revenue-neutral environmental tax reform actually introduced. By using a micro-simulation model we assess the distributional effects of this reform and found small effects both on households and environment.

Keywords: market-based instruments; air quality; distributional impacts; ETR; CEEC.

1. - Introduction

Since the late 1980s, there has been a growing appreciation of the potential of economic instruments for the internalization of environmental external costs. The role of price-based economic instruments gained momentum once an efficient response to environmental problems was sought (Goulder and Parry, 2008). The

* <*vojtech.maca@czp.cuni.cz*>; <*milan.scasny@czp.cuni.cz*>. This research was supported by Ministry of the Environment of the Czech Republic, R&D Grant No. SPII/4i1/52/07 MODEDR "Modelling of Environmental Tax Reform Impacts: The Czech ETR Stage II". The support is gratefully acknowledged. We are also grateful to Andrew Barton and Hana Škopková for their assistance in editing the manuscript. Responsibility for any errors remains with the authors.

idea of environmental taxation[1] levied on environmentally harmful products and activities has coincided not only with the endorsement of the polluter-pays principle,[2] but has become a key component of the long-standing debate on introducing a more efficient and fairer tax system and "greening" the budgets. Such a solution introduces so-called environmental tax reform (ETR) that basically shifts taxation from [economic] goods towards [environmental] "bads", such as pollution and resource use. In most cases, the ETR is considered revenue neutral; that is it should keep total tax revenues unchanged. The main argument for the ETR is about the possibility of reaping a double dividend, *i.e.* alongside with environmental quality improvement such a reform will boost employment and increase overall economic efficiency as well.[3]

We also document a similar tendency in the use of market-based instruments in environmental regulation in Central and Eastern European countries. During the early years of the transformation of their centrally-planned economies, state authorities in CEEC had to solve the problem of how to regulate economic agents that had to adapt to new conditions in order to improve fast the very bad quality of their environment. In spite of a few environmentally-related levies already in place (and many more were introduced in early 1990s), CEE governments tended to respond with stricter command-and-control instruments.

---

[1] Environmental tax is defined by OECD as «any compulsory, unrequited payment to general government levied on tax-bases deemed to be of particular environmental relevance» (OECD, 2001). However, in our study we consider environmental taxation in broader scope and consider any price-based economic instrument including emission charges, levies on a environmentally harmful product, or natural resource fee as "environmental taxation". We are therefore more in line with EUROSTAT (2001) that defines the environmental tax in such a broader scope as «...a tax whose tax base is a physical unit (or a proxy of it) of something that has a proven, specific negative impact on the environment».

[2] See *e.g.* OECD Recommendation of the Council on the Implementation of the Polluter-Pays Principle, C(74)223.

[3] Economic literature provides an argument only for existence of "weak double dividend", *i.e.* the ETR involves lower impacts on an economy if the additional revenues from taxation (or from auctioning of tradable rights) are recycled back to the economy *via* distortionary tax cuts rather than providing lump-sum. There is however no conclusive agreement about whether should the double dividend exist in general (the double dividend in its strong form). For further discussion of these issues see the paper by Markandya in this volume.

The first aim of this paper is thus to evaluate the use of market-based instruments introduced in ten countries of Central and Eastern Europe which are now member states of the European Union. We conclude that in most cases the market-based instruments introduced to regulate air quality and energy use are neither effective, nor correct the externalities. As they lack any motivation and allocation effect, they only fulfil a modest revenue rising objective by delivering the financial resources for specific environmental funds. The second aim of our paper is to an examine the possible effects of energy taxation introduced in a revenue-neutral fashion in one Central and Eastern European country – the Czech Republic.

Our paper is structured as follows. Section 2 summarises the workings of market-based instruments to regulate air quality and energy use in ten CEE countries (our study covers the following ten CEE countries: Bulgaria, Czech Republic, Estonia, Hungary, Latvia, Lithuania, Poland, Romania, Slovakia and Slovenia). In this section we also touch upon environmental tax reform plans discussed in some of these countries. Section 3 then examines the impacts involved in the implementation of Directive 2003/96/EC on energy taxation in the Czech Republic. The last section comprises the conclusion.

2. - Economic Instruments in Air and Climate Protection in CEE Countries

In the CEE region there is a long experience of using environmentally-related levies, well documented in two studies by the EEA (2000) and Regional Environmental Center (Speck *et* al., 2001). As evidenced in these reports, an excise tax on propellants and energy products already existed in the 1990s in each of the ten analysed CEE countries, and charges on NOx and SOx were also enforced in all but Slovenia (although Bulgaria, Hungary and Romania charged only an excess above the limit), and a CO_2 tax was introduced in Slovenia and Estonia.

In summary, the share of revenues from environmentally-related levies in CEE countries on total governmental revenues is broadly comparable with those in place in EU-15 countries (see Graph 1). Most of these revenues are generated by energy-related taxes, including excise taxes on propellants and transport taxes, while taxes on emissions and resource extraction typically constitute up to 5% of the revenues from environmentally-related levies only (Eurostat 2010 and EEA/OECD database).

GRAPH 1

ENVIRONMENTAL TAXES AS A PERCENTAGE OF GDP

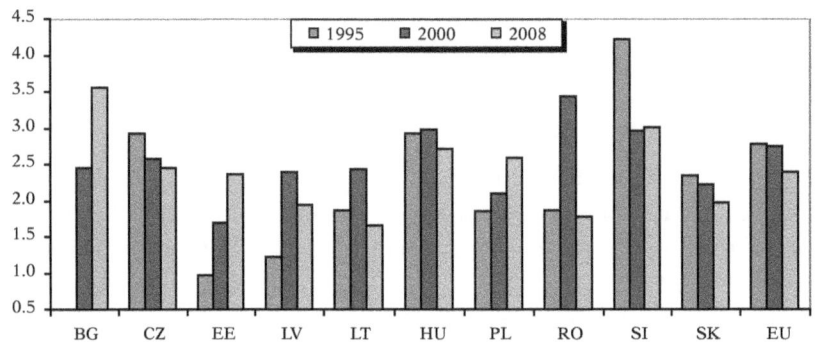

Source: EUROSTAT (2010).

Air Pollution Levies

At present, virtually all the CEE countries levy emission charges for a volume of pollutants emitted into the atmosphere. These levies are applicable irrespective of emission limits set for particular sources, although the standard rates are often multiplied by a certain factor in the case where they exceed an emission limit (Czech Republic, Estonia, Hungary, Lithuania, Poland and Slovakia). The emission charge rate can be even one order of magnitude larger than the standard rate in the case of non-compliance or illegal releases (*e.g.* Latvia, Poland, Romania). In contrast, in Bulgaria only the emissions of pollutants exceeding the limits are charged. In addition, some countries provide for a reduction of emission charge rates when abatement equipment is installed (*e.g.* Czech Republic, Hungary); see Table 1 for the charge rates and *Appendix* 1 for more details. Carbon dioxide is taxed less often and this tax exists recently only in Estonia, Slovenia and Poland.

TABLE 1

AIR POLLUTION CHARGES FOR STATIONARY SOURCES
(in EUR/ton as of 2008)

	SO_2	NO_x	Particulates/ dust	CO	CO_2	VOC/ organic carbon	Heavy metals
Bulgaria[*]	0-20	80-130	40-60			2,550	15,320 (lead)
Czech Rep.	40	32	120	24		80	801
Estonia	21	48	21	3	1.02	48	768
Hungary	199	477	107				
Latvia	42	42	5.6	7.5	0.1	18	1.155
Lithuania	83	138	53				348[**]
Poland	120	120	90-350	30	0.07	30	1,350-89,620
Romania	10	10		2.7	2.7	2.7	1,620-2,700
Slovakia	64	48	160	32		128	1,280/640[***]
Slovenia					0.01[†]		

Notes: [*]non-compliance fee, [**]class 1 substances, [***]class 1 / class 2 substances, [†]euro per emission coefficient.
Sources: OECD/EEA database on instruments used for environmental policy and natural resources management, national laws.

In terms of revenue volume, charges on airborne effluent pollutants are not an important instrument of regulation. For instance, the revenues from environmental charges for air pollution emissions, oil-shale mining and waste processing in Estonia amount to about €37m, the revenues from airborne pollutant charges in the Czech Republic amount to €15m, and non-compliance fees in Bulgaria generate annually less than one million Euro. Indeed, such an instrument lacks any motivation effect, which shows that any emission reduction has been driven rather by command-and-control instruments.

A part or all of the revenues from pollution charges are regularly earmarked for environmental funds for the financing of environmental projects in the majority of CEE countries. For instance, while in Poland pollution charges are divided among national, provincial, county and communal environmental funds in specified percentages that vary by effluent, only 20% of air pollution non-compliance fee revenues go to environmental funds in Bulgaria. In Czech Republic, Latvia, and Lithuania air pollution tax revenues are split between the National Environmental Fund and local (municipal) environmental funds to support environmental programmes.

With respect to the steering effect of environmental taxes and charges, a discrepancy between revenue raising and the pro-environmentally motivating objective still persists. In fact, very low rates of environmental taxes and charges provide quite limited incentives to achieve ambitious environmental goals. As Melichar *et al.* (2009) show, the air emission charge rates are several times below the marginal costs of pollution abatement and consequently provide little stimuli for further emission reductions.

GRAPH 2

COMPARISON OF NO$_x$ ABATEMENT COSTS
AND EMISSION CHARGE RATES
(in EUR/t)

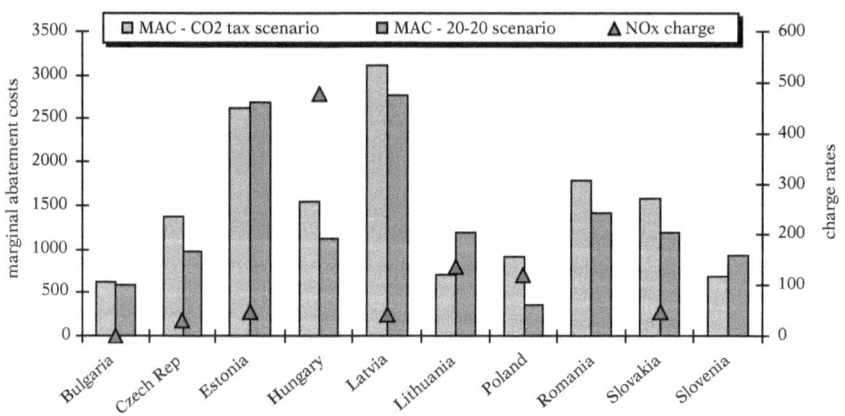

In spite of relatively differentiated systems of environmentally related charges (fees, taxes) in place, they have in common a lack of stimulating effects to achieve environmental policy targets, including the EU 20-20-20 goal and planned revision of national emission ceilings directive. The marginal abatement cost associated with the emission levels as set in the revised NEC Directive, assuming a moderate climate policy (the CO_2 tax scenario in Graph 2) or implementation of the EU climate/energy package for 2020 (20-20 scenario), is estimated by the CGE model GEM- E3 by Van Regemorter (2008), and these costs are as high as 1500 € per tonne of SO_2 or NO_x for the CO_2 tax scenario or about 680 € or 1300 € respectively for the 20-20 scenario (as averaged by us for our ten CEE countries). Except for Hungary and Poland, the actual rates of charges levied on these two pollutants thus correspond to only 2-6% of estimated marginal abatement costs (see Graph 2 for the details). Actual rates of charge levied on particulate matter are in some cases even three orders of magnitude lower than the relevant marginal abatement costs.

Moreover, existing rates of air emission charges are far removed from Pigovian rates. For instance, Zylicz (2002) argues that the emission charge rates in Poland were one order of magnitude lower than the Pigovian rate. Máca et al. (2010) draw a similar conclusion by comparing actual emission charge rates and the magnitudes of respective external costs as estimated by the ExternE method for several CEE countries. For instance, emission charges levied on stationary emission sources in Czech Republic internalise only about 0.5% of estimated damage.

Energy Taxation

The framework for energy taxation has been strengthened with Council Directive 2003/96/EC of 27th October 2003 which restructured the Community framework for the taxation of energy products and electricity and that entered into force in January 2004. Since CEE countries became members of the EU in May 2004 (except for Bulgaria and Romania, which joined in 2007) they were obliged to comply with fiscal structures and the levels of taxation to be imposed on energy products and electricity set in the directive.

However, even the minimum rates were deemed to create serious economic and social difficulties in view of the ongoing economic transition, relatively low income levels, comparatively low levels of excise duties previously applied and the limited ability to offset that additional tax burden by reducing other taxes. Consequently, various transitional arrangements were granted based on three principles:[4] strict limitation in time, proportionality to the objective addressed, and where applicable, with progressive alignment towards Community minimum rates. As of 2010, the majority of these transitional arrangements have already expired – which can be illustrated by comparing the excise tax rates applicable to motor fuels in 2005 and 2010 (see Graph 4 below).

GRAPH 3

IMPLICIT TAX RATE ON ENERGY
(in EUR/toe, deflated)

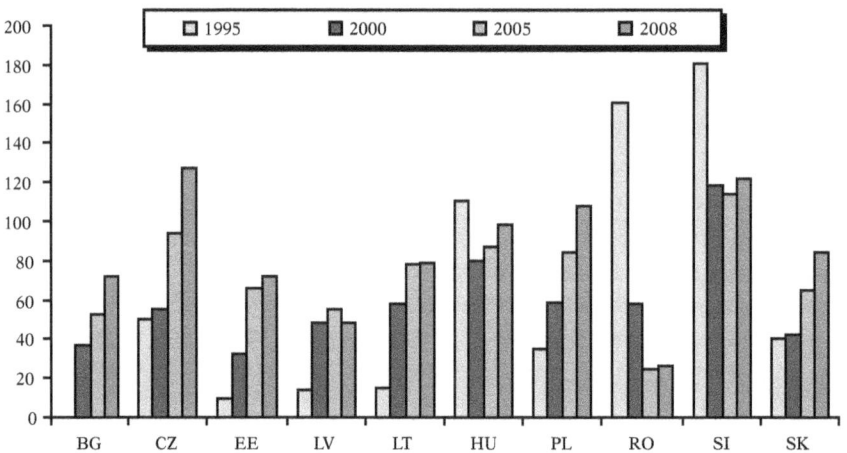

Source: EUROSTAT (2010).

---

[4] See Council Directive 2004/74/EC of 29th April 2004 amending Directive 2003/96/EC as regards the possibility for certain Member States to apply, in respect of energy products and electricity, temporary exemptions or reductions in the levels of taxation.

As we illustrate in the next two sections, the fear about the undesirable negative effects of energy taxes on the economy cannot be justified by model predictions. The rather minor economic impacts involved by the introduction of the Directive on energy taxation are mostly due to the fact that the rates of excise tax

GRAPH 4

EXCISE TAXES ON MOTOR FUELS
(in EUR per 1000 l)

DIESEL

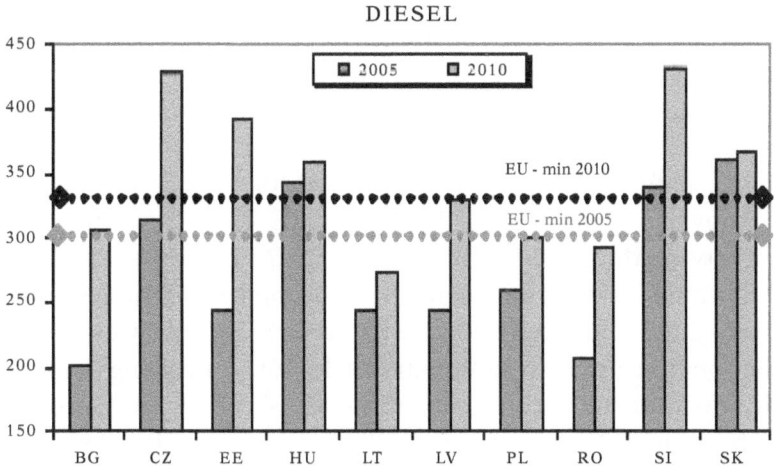

UNLEADED PETROL (95 RON)

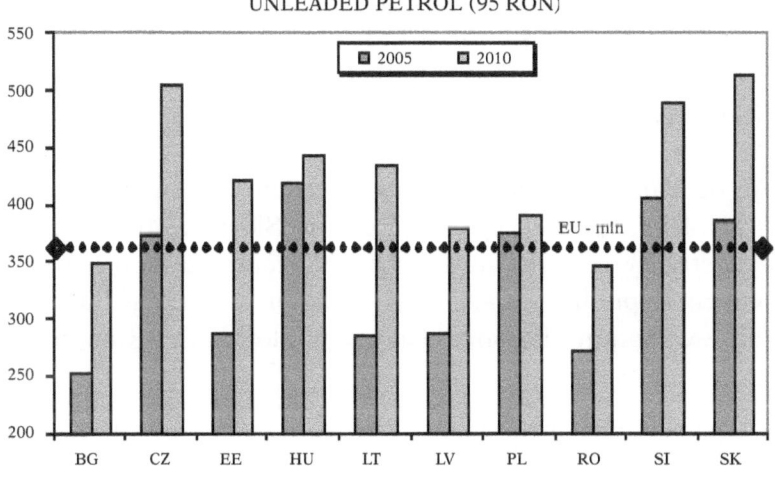

on propellants were already very close to the minimal level set by the directive in most of the CEE countries, and that taxes – especially on electricity – increased by only up to 1% of its final price only. However, a prevailing increasing trend in terms of implicit indirect tax rates on energy over the last few years is evident from the Graph 3 above.

Virtually all new member states were allowed an exemption for taxation of heating fuels and electricity, both a novelty in most CEE countries. Consequently, Poland still maintains a zero tax rate on coal for heating purposes (until 2012), while Slovakia endorsed a permanent tax exemption for coal used for heating in households. Natural gas is exempted in Bulgaria, Estonia and Lithuania according to Article 15(1)(g) of the Energy Taxation Directive (<15% share on total energy consumption), in Poland a transitional period was granted until 2013 and in Slovenia until May 2014. In Czech Republic natural gas is exempted if used for heating in households.

Hungary was perhaps the only CEE country with a tax on electricity at the time of EU accession. Both the Czech Republic and Slovakia introduced electricity taxation in 2007 (effective from January 2008 and July 2008, respectively). Estonia was granted a transitional regime until January 2010 for converting input electricity taxation into output taxation. In Slovakia, the tax was below the minimal rate even in 2009 but the rate doubled to EUR 1.32/MWh from January 2010. Bulgaria as well as Romania was later granted a transitional period as part of the Accession Treaty for the application of minimum excise duty rates for most energy products until January 2010. In fact, only Poland and Estonia tax electricity several times the Community minimum at 4.71 and 4.47 EUR per MWh respectively.

An optional tax exemption for electricity consumed in households has been adopted by Bulgaria and Slovakia. In some countries, electricity of renewable origin is exempted from electricity tax (Czech Republic, Poland), while other countries did not opt for this exemption (Estonia, Hungary, Slovakia).[5] Romania and

---

[5] It should be noted here that minimum levels of energy taxation set in the Directive has an unpleasant feature in that energy products for heat generation

Lithuania are the only CEE countries that differentiate between business and non-business taxation of electricity.

VAT on energy products and electricity is generally applied at a standard rate, except for delivery of natural gas and electricity to natural persons for private consumption in Latvia, and for firewood in the Czech Republic and Poland that is taxed at a reduced rate (10% and 7% respectively).

TABLE 2

VAT RATES APPLIED TO ENERGY PRODUCTS

	BG	CZ	EE	LV	LT	HU	PL	RO	SI	SK
Natural gas	20	20	20	10/21	21	25	22	24	20	19
Electricity	20	20	20	10/21	21	25	22	24	20	19
Firewood	20	10	20	21	21	25	7/22	24	20	19

Source: DG TAXUD (2010) VAT Rates Applied in the Member States of the European Union.

Emission Trading

Starting from 2005 CEE countries took part in the European Union's Emission Trading Scheme stipulated by Directive 2003/87/EC. However, the experiences from the first trading period (2005-2007) showed substantial over-allocation of emission allowances in all but one CEE country (Slovenia). Windfall profits gained from the sale of unused allowances by some energy producers point to considerable failures in the setting of allocation caps and may be seen as state aid. This was particularly evident when verified emissions for 2005 were published in May 2006 and the EUA price fell dramatically to under 10 EUR/ton.

---

are taxed as inputs while energy products for electricity generation are generally exempted and electricity is taxed as output. This compromise is not sufficient especially from environmental point of view as it precludes tax differentiation according to the environmental performance of fuels used for electricity generation except for tax exemptions and reductions allowed by the energy taxation directive that are however too narrow.

TABLE 3

EMISSION TRADING - ALLOCATION AND VERIFIED EMISSIONS

	no. of installations	annual cap 2005-2007 mil.t	difference between allocation and verified emissions (%) 2005-2007	annual cap 2008-12 mil.t
Bulgaria*		42.3		42.3
Czech Republic	414	96.9	15	86.8
Estonia	50	18.8	41	12.72
Hungary	254	30.2	15	26.9
Latvia	102	4.1	41	3.43
Lithuania	110	11.5	80	16.6
Poland	869	237.6	15	208.5
Romania	244	74.8	7	75.9
Slovakia	190	30.5	21	30.9
Slovenia	99	8.7	-2	8.3

Sources: EEA (2008), Community Independent Transaction Log (CITL).
Notes: Positive number in the difference between allocation and verified emissions indicates long position of given country, while the negatives indicate short position. * Bulgaria faced a significant delay in the preparation of the national allocation plan for the EU ETS for 2007 and was also the last member state to have its national allocation plan for the second trading period adopted by the European Commission.

Virtually all CEE countries also strongly opposed the negotiation of the recent amendment to the ETS Directive and especially the full auctioning of allowances, arguing that they have reduced CO_2 emissions during the 1990s more than the EU-15 and pushed for setting 1990 as a base-year rather than 2005. Finally, the French presidency put forward a compromise in the form of Article 10c that allows for free grandfathering of part of the allowances (70% in 2013 gradually decreasing to zero in 2020) in those countries that produced more than 30% of electricity from a single fossil source and where GDP *per capita* at market price was no more than 50% of the Community in 2006.[6]

---

[6] See Art. 10c (1) (c) of the Directive 2009/29/EC of the European Parliament and of the Council of 23th April 2009 amending Directive 2003/87/EC so as to

Environmental Tax Reform

Alongside discussions about the effective and efficient use of market-based instruments in environmental regulation, revenue-neutral environmental tax reform (ETR) has also been widely discussed since the beginning of 1990s. In the European Union, the political debate about the ETR was mostly linked to a discussion about possible revisions of energy taxation, especially when the so-called "Monti proposal" (COM(97) 30) and setting of new tax rates on diesel and petrol (COM(2002) 410) was on the agenda. Despite these long discussions, the concept of revenue-neutral environmental tax reform appeared at the EU level only in non-binding strategic documents (*e.g.* EC, 1993; 2007), leaving the decision on whether to increase energy taxes within or without a revenue neutral ETR fashion to each EU Member State (see 2003/96/EC). As a result, the ETR has been introduced in several "like minded" EU countries with different strengths and tax bases, and using a part of the revenues to support environmentally friendly programmes, and the revenue recycling fashion, if the revenue neutral principle was introduced at all (see OECD, 2001; 2006 for the review).[7]

In the CEE region, the concept of ETR started to be discussed widely from 2000, especially among NGOs and academia thanks to activities covered by the European-wide Platform on Environmental Fiscal Reform coordinated in that time by the European Environmental Bureau. For instance, the Hungarian Clean Air Action Group has been preparing an Alternative State Budget annually since 1999, and the Green Budget Working Group set up by the Environmental Committee of Parliament – the same as the Dutch Green Budget Commission – was established in 2001. In Slovenia, the ETR was announced along with introduction of a

---

improve and extend the greenhouse gas emission allowance trading scheme of the Community. All CEE countries but Slovenia fulfil this condition.

[7] The ETR was first introduced in Finland (1990) and Sweden (1991) with additional revenues recycled *via* income tax cuts, followed then by Norway and Denmark. In later years, complex tax reforms were introduced in the Netherlands (Energy Regulatory Tax in 2001), Germany (three-step increase in energy taxation 1999-2003), and the United Kingdom (a climate change levy with the use of revenues to promote energy saving measures introduced in 2001).

CO_2 tax in 1999 within the "Pinocchio project". The Estonian government discussed the preparation of an ETR concept several times after 2001 with the release of the ETR concept for public debate in 2005. On the other hand, no ETR was mentioned in Poland in official public documents at that time, although a working group on ETR was established by the secretary of State at the Polish Ministry of the Environment. In general, the stability of public finances, possible adverse impacts on the economy in transition and regressive impacts on households have been frequently mentioned as the main obstacles to the introduction of an ETR in the tax systems in most CEE countries.

This was also the case in the Czech Republic, where the ETR idea has been mulled over for quite some time in the political arena. Possibly the first ETR concept was drawn up by the Czech Ministry of Finance in 2000 (based on the "Monti proposal") but it was then decided to merge this proposal with a generally more comprehensive public finance reform prepared to stabilise the public finances. All of the proposals being prepared during next five years were then based on higher energy taxation, with the revenues used for labour income tax cuts, with a part of them earmarked for social compensation and investment in transportation infrastructure which contrasted with the revenue neutral principle originally claimed. The new centre-right government formed after the 2006 elections pledged to introduce a revenue-neutral ETR aimed at decreasing the energy intensity of the Czech economy and increasing employment. The coalition agreement anticipated a gradual implementation of an ETR – the first step required minimal excise taxes according to the energy taxation directive (2003/96/EC) to be introduced, and as part of the next two steps a new CO_2 tax should replace existing air pollution charges, the rates of energy taxes should be further increased and new vehicle tax should be introduced. So far, the first step of ETR has only been introduced, while governmental intention to introduce new CO_2 and vehicle taxes was abandoned. Most recently, a 10-fold increase of emission charge rates has been proposed in a new Air Protection Act draft that is meant to replace the CO_2 tax originally intended to be a key part of second phase of the Czech ETR.

A new impetus for ETR debate may be brought about with the revision of 2003/96/EC Directive and decision on auctioning of emission allowances for the third phase of the EU ETS. However, prospective revenues from both increased emission charging and auctioned EU allowances are proposed to be used to support environmental programmes and R&D technologies or to reduce governmental debt rather than to reduce labour taxation. Whether other CEE countries will follow the Czech case, or increase energy taxes – as proposed in the revision of 2003/96/EC Directive – within a revenue-neutral environmental tax reform will be seen soon.

3. - Assessment of the Latest Czech ETR Policy: the Implementation of 2003/96/EC Directive

During the transition, there were several cases when energy taxes were increased and labour or profit taxation was simultaneously cut. Although, such cases of tax change can classify as an ETR, until 2008 none of these change was ever supported explicitly by "green" rhetoric and referred to as an ETR (see Brůha and Ščasný, 2005*a* for the evaluation of tax policies). The only case of the ETR policy in the Czech Republic – introduced with the "green" rhetoric and with the revenue neutrality – was in fact the implementation of 2003/96/EC Directive as a part of the new government's economic policy package.

The Directive 2003/96/EC on taxation on energy products and electricity requires setting rate of relevant excise tax at a certain minimum level. With rates on diesel and petrol above the minimum rate at that time already, only a tax on solid fuels (8.50 CZK per GJ), a tax on natural gas (30.60 CZK per MWh) and a tax on electricity (28.30 CZK per MWh) had to be introduced in the Czech Republic.[8] These three taxes were adopted by Act no. 261/2007 Coll., on the stabilisation of public budgets and were enforced

---

[8] The equivalents of these rates are 1€ per MWh of electricity, 0.30€ per GJ of natural gas or solid fuels as set in the Directive.

from January 1st, 2008. While electricity is taxed on the output side, coal and natural gas are taxed on the input side, *i.e.* if consumed by households or power plants for heat generation. Use of gas for heating by households is however exempted from taxation at all.

It was foreseen that the revenues raised from these new environmental taxes will be offset by a corresponding cut in social security contributions paid by employers during the next year. Consequently, in late 2008, a cut in the rate of the state employment policy contribution by 0.4% has been adopted, effective as of January 2009. From a fiscal perspective, this cut broadly corresponded to the predicted revenues from the three aforementioned energy taxes.

In summary, we can conclude that this policy has had only a minor effect and in fact could not affect agent behaviour at all. The additional revenues from these three taxes are 3.2 bln. CZK in 2009 that corresponds to 0.22% of total public revenues, or 0.09% of GDP. We estimate the loss in revenues from the 0.4% cut in social security payments of 5.0 bln. CZK in 2009 that reduces overall tax burden, but in absolute terms only negligibly.[9]

The distributional impacts of this tax policy are evaluated by a micro-simulation model DASMOD (see *Appendix 2*). Due to the fact that centrally supplied heat is quite often consumed by Czech households living in apartment buildings, the second-order price effect of coal and gas taxation on centrally supplied heat is included in our impact evaluation. General equilibrium effects are however not considered.

Overall, energy taxes as introduced in 2007 in the Czech Republic (EC07 hereinafter) do not present a dramatic change in residential energy prices. In fact, the final price of energy is predicted to increase by approximately 11% for solid fuels and 1%

---

[9] *Ex ante* estimate of additional revenue by the study by ŠČASNÝ M. and BRŮHA J. (2007) for the scenario with natural gas exempted from the taxation was of 3.7 bl. CZK, while the Czech Ministry of Finance estimated the revenues as high as 4.3 bln. CZK, Most likely, this overestimation is responsible for setting the rate of social security contributions at the level that generates larger losses than the additional revenues.

for electricity. The price of gas remains unchanged thanks to the exemption provided from taxation. Price of heat raises due to the second-order effect by 1.5% (Ščasný and Brůha, 2007). As a result, consumption of solid fuels and electricity is predicted to fall by about 1%, consumption of gas remains unchanged, while heat use decreases by 2.6% (see A1 in Table 4). The governmental decision about exempting natural gas used in households for heating purposes leads to an increase of gas and electricity consumption and, thanks to cross-price effects, to a reduction in heat use (compare A1 and A2 in the table). While the direct (the first order) effects on the price are only considered, one could incorrectly draw conclusions about the impact on energy carriers use; for instance, consumption of gas by households is reduced, while it actually increases due to the cross-price effects (compare A1 and B1). Overall, the effect of the energy taxation on the consumption of four analysed energy types is however quite small.

TABLE 4

PREDICTION OF ENERGY CONSUMPTION FOR EC07
WITH ALTERNATIVE POLICY SETTING

	solid fuels	gas	heat
A1. EC07 exemption on gas (heat II-order effect)	-1.10%	0.14%	-2.64%
A2. EC07 with gas tax (heat II-order effect)	-1.09%	-3.65%	-0.14%
B1. EC07 exemption on gas (heat I-order effect)	-1.11%	-0.36%	-1.28%
B2. EC07 with gas tax (heat I-order effect)	-1.09%	-4.13%	1.26%

Similarly, the aggregated impact on expenditures on energies, welfare and public revenues is very small (see Table 5). For instance, additional tax revenues from household taxation are estimated at CZK 1.2 billion (or 45 million €) annually, which corresponds to less than 0.1% of total public revenues, or 0.04% of GDP. Keeping revenue neutrality, the SSC rate can be reduced by 0.18%-points, or the lowest PIT rate by 0.28%-points, or tax credits of CZK 7,200 per taxpayer and year can be increased by CZK 312.[10] Aggregated impacts for scenarios without the price effect on heat price or including the taxation on natural gas are also reported in the table, while the results for EC07 are shaded.

TABLE 5

PREDICTIONS OF AGGREGATED IMPACTS
FROM THE TAXATION OF HOUSEHOLDS

	energy expenditures (bln. CZK)	welfare (bln. CZK)	public revenues (bln. CZK)	potential for labor taxation cuts
with the exemption on natural gas				
with the price effect on price of heat				
EC07 without recycling	0.01	-1.29	1.19	
EC07 SSC (12.5%)	0.09	0.19	1.22	-0.18%
EC07 CREDIT (7.200 CZK)	0.10	0.19	1.22	312 CZK
EC07 PIT (12.0%)	0.10	0.19	1,22	-0.29%
without the price effect on price of heat				
EC07 without recycling	0.24	-0.85	0.81	
EC07 SSC (12.5%)	0.30	0.12	0.83	-0.12%
EC07 CREDIT (7.200 CZK)	0.30	0.12	0.83	202 CZK
EC07 PIT (12.0%)	0.30	0.12	0.83	-0.19%
including taxation on natural gas				
with the price effect on price of heat				
EC07 without recycling	0.47	-2.34	2.06	
EC07 SSC (12.5%)	0.61	0.14	2.11	-0.30%
EC07 CREDIT (7.200 CZK)	0.62	0.15	2.11	521 CZK
EC07 PIT (12.0%)	0.62	0.15	2.11	-0.48%

---

[10] Obviously, the assumption used for household responses in the microsimulation model also has an effect on the model's predictions. This is documented by an experiment by Ščasný M. and Brůha J. (2007) who assume three values of elasticities in their simulation of ETR2007 incidence; first, they use the elasticities estimated for specific household types; second, they use the average of these values; and lastly, they assume no response, *i.e.* zero elasticities. They further show that to ensure, for instance, revenue neutrality for the EC07 scenario, an authority

The effect on energy expenditures and welfare are predicted to be very small on average, and it is less than 0.1% of total household expenditures. These effects for various household groups amount at most to some hundreds of CZK. Energy expenditures are increased mostly in households living in smaller cities, and in households using solid fuels for heating (COAL). On the other hand, households connected to a centralized heating system would have slightly lower expenditures on energy. Revenue recycling switches welfare loses into a welfare gain in all household groups. The only losers for all four EC07 scenarios are households of retired people. Table 6 displays the distribution of welfare and expenditure impact in detail for several groups of households.

Macroeconomic effects were not analysed *ex ante* for the EC2007 policy, during the time of Directive implementation. However, Ščasný and Brůha (2007) estimate the first order effect on energy expenditures and consumption of economic sectors. To properly consider behavioural responses of firms, they analyse changes in the factor composition involved by predicted price increases. Specifically, they estimate elasticities of substitution for four types of energy, labour and capital from the nested production function (Berndt and Wood, 1979) separately specified for several economic sectors of the Czech economy. They conclude that overall first-order effects are relatively small because the tax rate increase is small and because coal and gas use for electricity production and heat generation in co-generation are excluded from taxation. Except for a few sectors such as the manufacture of leather products, energy expenses are predicted to increase by some percentage units. To support previous findings, Zimmermannová (2009) constructed a simple price input-output model and found that the effect of EC2007 on final price is almost negligible, about 0.2% in absolute terms, in most of the economic sectors.

---

should increase PIT tax credits by CZK 407 per year and employee, or by CZK 419 or CZK 423, respectively, for the next two assumptions on the magnitude of the demand elasticities. For their specific case, the prediction model would thus overestimate the labour tax cuts by 3%, or 4% for the averaged or zero (incorrect) elasticities, respectively.

TABLE 6

PREDICTIONS ON HOUSEHOLD IMPACTS FOR THE EC07 POLICY SCENARIOS

	EC07				EC07-PIT				EC07-SSC				EC07-CREDIT			
	energy expenses	energy taxes	labour taxes	welfare	energy expenses	energy taxes	labour taxes	welfare	energy expenses	energy taxes	labour taxes	welfare	energy expenses	energy taxes	labour taxes	welfare
1	37	296	0	-319	65	306	-418	99	55	302	-286	-33	64	306	-396	77
2	40	301	0	-324	63	309	-367	43	58	307	-286	-38	62	309	-339	15
3	-9	285	0	-310	8	291	-265	-45	6	290	-236	-74	7	291	-254	-55
4	3	281	0	-305	16	285	-212	-94	15	285	-202	-103	15	285	-205	-101
5	-11	271	0	-297	2	276	-238	-59	1	275	-229	-68	3	276	-238	-59
6	13	283	0	-306	29	288	-266	-39	27	288	-258	-48	29	289	-269	-36
7	16	268	0	-292	38	276	-374	82	36	276	-374	81	39	277	-395	103
8	-6	292	0	-319	20	301	-441	123	19	301	-457	139	21	301	-461	142
9	-8	281	0	-307	19	291	-488	181	20	291	-538	231	20	291	-504	197
10	-55	273	0	-303	-31	282	-475	171	-22	286	-673	370	-30	282	-482	179
ELECTR	114	216	0	-235	140	226	-368	133	136	224	-340	105	140	226	-365	129
ELEcookGAS	56	114	0	-125	71	118	-180	55	63	116	-146	21	72	118	-206	81
HEATcookELE	-285	258	0	-310	-265	264	-318	8	-266	264	-326	16	-265	264	-316	6
HEATcookGAS	-191	278	0	-314	-173	284	-342	29	-173	284	-360	46	-172	284	-344	30
GAS	30	148	0	-168	50	156	-367	199	48	155	-361	193	50	156	-367	200
SOLID	903	913	0	-886	933	925	-395	-491	928	923	-357	-528	932	925	-387	-498
farmer_small	385	458	0	-459	415	470	-487	28	413	469	-465	7	416	470	-498	40
farmer_large	158	361	0	-379	193	374	-495	116	191	373	-479	100	192	373	-482	102
retired_small	318	423	0	-430	318	423	0	-430	319	423	-8	-423	318	423	0	-430
retired_middle	-4	207	0	-227	-4	207	0	-227	-3	207	-7	-219	-4	207	0	-227
retired_large	-102	199	0	-224	-102	199	0	-224	-101	199	-8	-216	-102	199	0	-224
EA1_small	127	227	0	-239	149	235	-282	43	142	232	-230	-9	152	236	-315	76
EA1+_small	241	371	0	-382	269	382	-399	17	261	379	-301	-82	264	380	-323	-60
EA2_small	249	385	0	-399	286	400	-583	184	287	401	-611	212	289	401	-628	229
EA2+_small	404	511	0	-515	447	528	-628	113	444	527	-608	92	449	529	-660	145
EA1_large	-67	275	0	-307	-45	283	-386	79	-47	282	-373	66	-45	283	-384	77
EA1+_large	-83	269	0	-302	-60	277	-395	94	-65	275	-346	44	-64	276	-335	33
EA2_large	-54	286	0	-318	-23	298	-587	269	-20	300	-659	341	-21	299	-622	304
EA2+_large	-82	311	0	-348	-45	324	-652	304	-47	324	-665	316	-44	325	-673	324

It is not surprising that the effect on the consumption of electricity and solid fuels is less than 1%, although the effect on the consumption of natural gas is quite larger thanks to the larger price responsiveness of gas demand as estimated by Ščasný and Brůha (2007). Additional revenues are estimated of less than CZK 3 billion yearly (100 million Euro) that could be used to reduce SSC paid by employers – keeping revenue neutrality – from 35% by only 0.04%. Without doubt, this policy has a negligible effect on the employment as well.

4. - Concluding Remarks

Market-based instruments are used in CEE countries to regulate air quality quite widely and from the scattered picture of the application of economic instruments it is evident that these countries have an abundant number of air pollution charges and other market-based instruments in place. The environmentally-related charges and fees have been introduced with rates that are, however, neither effective, nor efficient. The rates of emission charges thus barely provide any incentive for air pollution abatement. Together with energy taxes, mostly introduced as excise tax, they serve rather as revenue-raising instruments for dedicated purposes (such as support of transportation infrastructure build-up), to contribute to environmental funds or just bring in revenue to finance public policies.

The large number of airborne pollutants taxed and taxpayers also points to likely inefficiencies, *e.g.* a recent study in Cezch Republic revealed that administrative costs of air pollution charges to small and medium sources are about 150% of the revenues collected (Pavel, Vítek, 2009). In some CEE countries, discussions about increasing the effectiveness of the regulation in air and climate protection are now taking place, aiming at making best use of market-based instruments, among others.

Without any exception, CEE countries are also faced with inefficiencies due to concurring regulation – overlapping between various market-based instruments and with command-and-control instruments that effectively lead to double or triple burdening. Yet this debate is unresolved even at the EU level especially with respect to integrated pollution prevention and control (IPPC) and emission trading. A thorough assessment of the IPPC directive,[11] carried out in the course of recast of the directive, identified only a limited option for the use of market-based instruments under the existing IPPC regime. Firstly, the IPPC provides no incentive for dynamic efficiency – unless a new permit is needed (or requested by a public authority) the operator is not motivated to maintain advancements in BATs. Secondly, each installation is required to comply with BAT-based emission limits set in each individual permit thus provides very limited room for quantity-based economic instruments (*i.e.* emission trading).

The double burdening argument has also been raised with respect to taxation of energy products used by installations covered by the EU ETS. While this debate is nowadays more pronounced in old EU member states (especially those with higher energy taxes), it will likely gain more attention with the tightening of caps under the EU ETS and the full auctioning of emission allowances. The position of the European Commission is elaborate in this respect, arguing that the objective of the Energy Taxation Directive is not only environmental protection but at the same time harmonization of excise taxes, which contributes to the proper functioning of the Single Market. To make it more complicated, the Commission suggests that installations with the best performing techniques may still be granted full tax exemptions[12], thus effectively pointing back to IPPC-derived standards.

The introduction of environmental tax reform made an appearance in political debate especially after the political and economic transformation, although without any real effect. As documented in the case of Czech Republic, although several specific ETR proposals have been discussed since 2000 by the government, the only real revenue-neutral ETR policy was introduced in 2007

---

[11] See SEC(2007) 1679, p. 99.

[12] See Community guidelines on State aid for environmental protection, OJ C 82, 1. 4. 2008, and Commission Decision on aid scheme C 41/06 (Denmark CO_2 tax scheme), OJ L 345, 23. 12. 2009.

as a response to the implementation of 2003/96/EC Directive on taxation on energy products and electricity. However, this policy does not have any significant effect on the situation of households, nor on the economy as shown in the previous chapter.

In general, in most of the CEE countries, the two main reservations about the ETR have been frequently encountered in the political debate. The first relates to the question whether the ETR will boost employment at all rather than having a harmful impact on the competitiveness and growth of the economy. The second is centred around the possible regressivity of an energy tax and the corresponding undesirable impact on certain households, like the retired, jobless, or single mothers in particular.

Telling whether a green tax shift has a harmful effect on the economy and on the employment is indeed not a trivial question to answer without relying on a macro structural model. Both, general equilibrium based and macro-econometric models can in fact serve as a very useful tool for the prediction of a "what if" policy scenario. Having them at hand, the authorities can get a better picture of the possible adverse medium- and long-term effects of proposed policies, allowing them to adjust the design of proposed instruments or to introduce measures that mitigate potential adverse effects. In spite of this fact, there is much less experience in the use of the macro structural models for an impact assessment of environmental policies or energy taxation in CEE countries compared to Western Europe, especially during 1990s (see Czajkowski *et al.*, 2010, or Ščasný *et al.*, 2009 for a review). According to our best knowledge, in most cases, the incidence of policies has been often assessed by using technology-based engineering optimisation models. However, this type of model usually lacks interactions and feedbacks with the reminder of the economy with overall consumption and/or output growth exogenously predefined, and as such unsuitable to assess any impact on the economy. This fact underscores the need for developing environmentally-extended macroeconomic models in CEE countries. Low quality and unavailability of statistical data, that hindered development of these models especially during 1990s, have both substantially improved in latest years making such modelling more feasible. In recent years, it results in development several CGE models in Czech Republic (*e.g.* Kiulia, 2009) or Poland (Czajkowski *et* al., 2010), or an extension of environmental and energy module of the macro-econometric E3ME model (see Barker et al., 2007; Pollitt, 2008 for the model description) and its application for policy evaluation purposes in Czech Republic (*e.g.* Šťasný *et* al., 2009).

Determining regressivity of energy taxes and undesirable adverse distributional and social effects can be only evaluated by a micro-simulation model, the best being properly embodied by parameters on price responsiveness of households. Such evaluation for the case of the incidence measurement of energy taxes as introduced during implementation of 2003/96/EC Directive in Czech Republic has just been presented. Thanks to such model, one can identify winners and losers of intended environmentally-friendly policy while considering also various options for the revenue-recycling and/or provision of an income-tested or a lump-sum transfer. Specifically, we evaluate the impacts on household expenditures and welfare that we believe both provide useful information in a decision-making. While the prediction about the impact on household expenditures may inform policy makers about expected fiscal impacts, environmental effects, but also about the possible targeting of social mitigation measures and support mechanism to enhance energy saving installations in households, the prediction about induced welfare loss or benefit inform policy maker about economic (Pareto) efficiency and the desirability of proposed policy. Such exercise requires however quite reach micro-level dataset and properly estimated household demand.

In the second stage of the Czech ETR policy, the rates of air emission charges are suggested to be increased ten-fold in a newly drafted air quality law, alongside with a substantial reduction in the number of taxed pollutants. One should be cautions to become optimist; although this proposal was drafted in 2009, it has not reached parliament yet. Moreover, based on the prediction by E3ME macro-econometric model, Šťasný and Píša (2009) predict that the impacts of this policy on the economy, the employment and the environment will be negligible. Some additional revenues for the environmental fund remain to be the only real effect of this policy. If the revision of 2003/96/EC Directive on taxation on energy products and on electricity, or introducing the allocation of a part of allowances *via* auctions in the third phase of EU ETS brings new impetus for more efficient regulation in CEE countries will be seen in the near future. Whether these policies result in adverse economic and/or distributional impacts can be evaluated *ex ante* only if appropriate simulation models are at hand.

Air Pollution Levies in CEE Countries

The rate of air pollution non-compliance fee in Bulgaria is related to the release of emissions above the environmental standard and in correspondence with the total volume of pollution based on the quantity of working hours of the respective facility. There are different rates set for 16 different polluting substances, including NO_2, lead, phenols, SO_2 and dust. Non-compliance fees are differentiated also in correspondence with the geographical location. For instance, rates are tripled if the pollution takes place near national parks and doubled if it takes place near protected regions, water-supply sources or sanitary zones.

Estonia has been using environmental taxes and charges for air and water pollution since 1991. The fees were continuously increased by around 10-20 per cent annually since 1996. For non-compliant polluters a basic fee is multiplied by 10 (for exceeding limits) or by 20 (operation without permit). Unlike the majority of new Member States Estonia charges an environmental fee for CO_2 emissions.

In Hungary an environmental load charge has been introduced in 2003 covering pollution released to air, water and soil. The charge is paid by installations subjected to permits and does not apply to households. The charge was introduced gradually, so 40 per cent of the calculated charge was applied in 2004 and 2005, 75 per cent in 2006, 90 per cent in 2007 and 100 per cent in 2008 and onwards. An extra penalty is charged for the exceeding of air quality regulation. Yet, there is a possibility of 50 per cent reduction in case of commencement of abatement equipment installation.

In Latvia the charge for use of natural resources is *inter alia* levied on the volume of greenhouse gasses emitted by installations, which are not covered by greenhouse gas emission quotas (0.1 LVL/t). In addition, the tax is also applied on emissions of solid, carbon monoxide, ammonia and other non-organic compounds. A non-compliance fee is set at 3 times the charge for the respective pollutant, while for illegal emissions and non-reporting the rate is 12 times the charge.

LITHUANIAN tax on pollution was first promulgated in 1991. The tax is imposed on both stationary and mobile sources (including aircrafts). The tax is levied on emissions of SO_2, NO_x, solid particles and four groups of other pollutants. The emissions above the permitted level are charged by a penalty rate.

Also POLAND has a two-tier environmental tax structure. Firms are required to pay pollution charges for emissions below an administratively specified limit, and additionally, usually much higher pollution fines for emissions exceeding the specified limit. Currently, there are 67 of air polluting gases and particulate matters whose emissions are charged for. In addition, there are 7 types of gases that are emitted during transhipment of motor fuels. Some 16 gases, emitted from small coal, oil, gas or wood-fired boilers that do not require license and constant emission monitoring are taxed depending on the boiler type and size. There are 32 gases and particulate matters originating from internal combustion engines that are taxed depending on the engine type and size. In total, 16 air pollutants emitted from poultry farms are taxed depending on stock size and farm type. Environmental fines for exceeding emission limits or illegal release of emissions are charged at ten times the rate for the respective "regular" environmental charge.

Air pollution charges were among the first economic instruments introduced in CZECH REPUBLIC (or Czechoslovakia those days) in 1967. Currently, the Air Protection Act of 2002 sets the detailed rules on charging operators of stationary air pollution sources basically according to installed capacity. The charges (except for small pollution sources below 200 kW) are due for a group of core pollutants including particulate matters, sulphur dioxide, nitrogen oxides, volatile organic compounds, polycyclic aromatic hydrocarbons (PAHs) and heavy metals, and two classes of other harmful substances (*e.g.* benzene and its compounds). Until 2007, the taxation of natural gas was a part of excise duty on mineral oils and except for use as propellant a zero tax rate applied. Tax base is the amount of gas in MWh of gross heating value and the tax rates are set according to type of use. In principle two tax rates are set – CZK 264.8 per MWh (EUR 2.894 per GJ) for use as propellant and CZK 30.6 per MWh (EUR 0.334 per GJ) for heat production, stationary engines and construction works.[13] In order to promote natural gas as automotive fuel there is a special tax reduction arrangement – until 2012 zero tax rate applies and in following years it will gradually rise up to CZK 264.8 per MWh in 2020. This tax reduction is a complementary measure to a voluntary agreement between government and gas industry on promotion of natural gas as motor fuel.[14] The compliance with energy taxation directive is based on Article 15 (1) *(i)* that allows for total or partial exemption under fiscal control for natural gas used as propellant. The tax on solid fuels is an entirely new tax, until now solid fuels were subject only to VAT. Enumeration of taxed commodities includes hard coal, lignite, briquettes, coke and other solid fuels used for a production of heat. The tax base is the amount of solid fuels in GJ of gross calorific value with only one tax rate of EUR 0.334 per GJ.[15] Calorific value of a fuel sample shall be proven by an accredited laboratory. If such measurement is not available a default value of 33 GJ per ton will be used. Tax exemptions are notably granted to solid fuels used for electricity generation and combined heat and power generation in highly-efficient generators if the heat is delivered to households. The electricity tax is calculated as the amount of electricity in MWh consumed and the tax rate is CZK 28.3 per MWh (EUR 1.113 per MWh)[16] regardless whether the electricity is used for business or non-business purposes. There are some tax exemptions put in place – foremost for environmentally friendly elec-

---

[13] The minimal rates prescribed by the directive are EUR 2.6 per GJ and EUR 0.3 per GJ applicable to motor fuels and heating fuels respectively.

[14] See also Government resolution of 11th June 2005 no. 563 on Program for promotion of alternative fuels in transport - natural gas.

[15] The minimal tax rate set by the energy taxation directive is EUR 0.15 per GJ for business and EUR 0.3 per GJ for non-business use.

[16] The minimal tax rate set in the energy taxation directive is EUR 0.5 per MWh for business and EUR 1 per MWh for non-business use.

tricity – that is defined as electricity of solar, wind or geothermal origin, produced in hydroelectric installations, generated from biomass or from products originating from biomass, generated from methane emitted by abandoned coal-mines or generated in fuel cells.

APPENDIX 2

Assessment of Distributional Effects by the Micro-simulation Model DASMOD

Distributional effects of energy taxation are assessed by the DASMOD static micro-simulation model. This model allows the prediction of changes of energy prices on household expenditures and consumption, welfare of households, and public revenues. Utilizing information about predicted changes on energy consumption, the model also computes changes in environmental damage by calculating the external costs associated with energy generation or fuel use. See Brůha and Ščasný (2006; 2008) for the detailed description.

The DASMOD model also allows for simulations of changes in labour taxation including the revenue neutral tax reform. Specifically, the model allows simulating the revenue recycling *via* personal income tax rate(s) cuts, changes in tax deductibles or provision of tax credit, or changes in social security contributions obligatory paid by employers or employees. Social transfers – income tested or not – provided to mitigate adverse social effect involved by a policy can be also a part of parametrical changes of the model.

Due to the fact that a consumer may response to price and income changes, the model counts for the price and income elasticities that were estimated from the coherent demand system, namely *Almost Ideal Demand System, AIDS* (Deaton and Muelbauer 1980). To consider heterogeneity among household groups, the DASMOD model contains several sets of elasticities that were estimated separately in earlier empirical work for several groups of households (see Brůha and Ščasný, 2005*b*; 2006).

The model is built on micro-level data from the Czech Family Budget Survey regularly conducted by the Czech Statistical Office. Predictions of changes of all endogenous variables are always performed at the lowest possible level, *i.e.* we predict changes for each of almost 3000 households. The predictions for each individual household are then clustered according to pre-defined household groups and segments such as deciles and aggregated for the entire Czech population by using PKOEF variable.

BIBLIOGRAPHY

BARKER T. - JUNANKAR S. - POLLITT H. - SUMMERTON P., «Carbon Leakage from Unilateral Environmental Tax Reforms in Europe, 1995-2005», *Energy Policy*, no. (35), 2007, pages 6281-6292.

BERNDT E. - WOOD D., «Engineering and Econometric Interpretations of Energy-Capital Complementarity», *The American Economic Review*, June 1979, no. 69(3), 1979, pages 342-354.

BRŮHA J. - ŠČASNÝ M., «Environmental Tax Reform Options and Designs for the Czech Republic: Policy and Economic Analysis», in DEKETELAERE K. - KREISER L. - MILNE J. - ASHIABOR H. - CAVALIERE A. (eds.), *Critical Issues in Environmental Taxation: International and Comparative Perspectives*, Vol. III, Richmond Law & Tax Publisher Ltd., Richmond, UK, March 2005, pages 750, Chapter 17, 2005*a*.

— — - — —, «Analýza distribučni efektů regulace v oblasti spotřeby energií a dopravy», *Working Paper*, připravený pro Závěrečnou zprávu projektu VaV MŽP 1C/4/73/04, Environmentální a ekonomické dopady ekonomických nástrojů envionemtnální regulace, Centrum pro otázky životního prostředí Univerzita Karlova v Praze, 2005*b*.

— — - — —, *Distributional Effects of Environmental Regulation in the Czech Republic*, Paper presented at the 3rd Annual Congress of Association of Environmental and Resource Economics AERE, Kyoto, 4-7 July, 2006.

— — - — —, *Distributional Effects of Environmentally-Related Taxes: Empirical Applications for the Czech Republic*, Paper presented at the 16th Annual Conference of the European Association of Environmental and Resource Economists, Gothenburg, 25-28 June 2008.

CZAJKOWSKI M. - HAGEMEJER J. - KIUILA O. - PISA V. - ŠČASNÝ M. - ŻYLICZ T., *Factors of Efficient Internalization in CEE Counties and Policy Recommendations*, Report of the FP6 EXIOPOL project, 2010.

DEATON A.S. - MUELLBAUER J., «An Almost Ideal Demand System», *American Economic Review*, no. 70, 1980, pages 312-336.

DG TAXUD, *VAT Rates Applied in the Member States of the European Union*, Brussels, European Union, 2010.

EEA, *Application of the Emissions Trading Directive by EU Member States - Reporting year 2008*, Technical Report 13/2008, Copenhagen, EEA, 2008.

EUROPEAN COMMISSION, *White Paper on Growth, Competitiveness, Employment*, 1993.

— —, *Green Paper on Market-Based Instruments for Environment and Related Policy Purposes*, [SEC(2007) 388] /COM/2007/0140 final/, 2007.

EUROSTAT, *Environmental Taxes - A Statistical Guide*, European Commission, Theme 2, Economy and Finance, Office of Official Publication of the European Communities, Luxembourg, 2001.

— —, *Taxation Trends in the European Union*, Office of Official Publication of the European Communities, Luxembourg, 2010.

GOULDER L.H. - PARRY I.W.H., «Instrument Choice in Environmental Policy», *RFF, Discussion Paper*, Resource For The Future, Washington DC, 2008.

KIULIA O., «Computable General Equilibrium Model for the Czech Economy», in ŠČASNÝ M. (ed.), *Modeling Of Environmental Tax Reform Impacts: The Czech ETR Stage II*, The 2009 Annual Report of SPII/4i1/52/07 R&D Project, Chapter 6.1, Charles University Environment Center, Prague and Institute of Economic and Environmental Policy, University of Economics Prague, 2009.

Máca V. - Melichar J. - Ščasný M., «External Costs From Energy Generation and Their Internalisation in New Member States», in Soares C.D. - Milne J. - Ashiabor H. - Deketelaere K. - Kreiser L. (eds.), *Critical Issuess in Environmental Taxation*, Vol. VIII, Oxford University Press, Oxford (forthcoming), 2010.

Melichar J. - Máca V. - Ščasný M., «Environmental Externalities, Abatement Costs and Market Based Instruments», in Ščasný M. (ed.), *Modeling Of Environmental Tax Reform Impacts: The Czech ETR Stage II*, The 2009 Annual Report of SPII/4i1/52/07 R&D Project, Chapter 3, Charles University Environment Center, Prague and Institute of Economic and Environmental Policy, University of Economics Prague, 2009.

OECD, *Environmentally Related Taxes in OECD Countries*, Issues and Strategies, Paris, 2001.

— —, *The Political Economy of Environmentally Related Taxes*, Paris, 2006.

Pavel J. - Vítek L., «Administrative Costs of the Czech System of Environmental Charges», in *Kol. Critical Issues in Environmental Taxation*. Oxford, Oxford University Press, ISBN 978-0-19-956648-8, 2009, pages 247-261, 984 s.

Pollitt H., *E3ME: An Energy-Environment-Economy Model for Europe A Non-Tech nical Description - version 4.6*, Cambridge Econometrics, Cambridge, 2008.

Ščasný M. - Brůha J., *Predikce sociálních a ekonomických dopadů návrhu první fáze ekologické daňové reformy České republiky (Prediction of Social And Economic Impacts of the First Phase of Environemntal Tax Reform in the Czech Republic)*, a study prepared for the Czech Ministry of the Environment, Charles University Environment Center, Prague, April 2007.

Ščasný M. - Píša V., «Dopady změn zpoplatnění emisí znečisťujících látek vypouštěných do ovzduší ekonometrickým modelem E3ME», in Ščasný M. (ed.), *Modeling Of Environmental Tax Reform Impacts: The Czech ETR Stage II*, The 2009 Annual Report of SPII/4i1/52/07 R&D Project, Chapter 5, Charles University Environment Center, Prague and Institute of Economic and Environmental Policy, University of Economics Prague, 2009.

Ščasný M. - Pollitt H. - Chewpreecha U. - Píša V., «Analysing Macroeconomic Effects of Energy Taxation by Econometric E3ME Model», *Czech Journal of Economics and Finance*, no. 5/2009, Vol. 59, 2009, pages 460-491.

Speck S. - McNicholas J. - Markovic M., *Environmental Taxes in an Enlarged Europe: An Analysis and Database of Environmental Taxes and Charges in Central and Eastern Europe*, The Regional Environmental Center for Central and Eastern Europe, Szentendre, 2001.

Van Regemorter D., *Assessment of the Macroeconomic Impacts of NEC Scenarios with GEM-E3*, April 2008.

Zimmermannová J., «Dopady zdanění elektřiny, zemního plynu a pevných paliv na odvětví OKEČ v České republice», *Politická Ekonomie*, Vol. 57/2, 2009, pp. 213-231.

Zylicz T., «Může existovat "ekologická daň"?», in Ščasný M. (ed.) *Konsolidace vládnutí a podnikání v ČR a v Evropské unii*, část 5, Praha, Matfyzpress, 2002.

II - DISTRIBUTIONAL ISSUES

Environmental Fiscal Reform in East and Southern Africa and its Effects on Income Distribution

Daniel Slunge - Thomas Sterner*

University of Gothenburg

The paper reviews the current use of instruments for environmental fiscal reform (EFR) in selected East and Southern African countries and analyzes the effects on income distribution from fuel taxes. Theoretical arguments for introducing taxes on environmental and fiscal grounds as well as potential trade-offs between environmental and fiscal objectives are discussed. While most African countries have introduced several environmental taxes, our analysis indicates there is a considerable potential to improve both revenue generation and environmental benefits. Building on detailed case studies of fuel consumption, we find that fuel taxes appear to be progressive and not regressive as often claimed. [JEL Classification: Q56, Q58, H23, H22].

Keywords: environment; distributional effects; tax revenue; fuel taxes.

1. - Introduction

Environmental fiscal reform (EFR) encompasses a range of taxation and pricing instruments on environmental pollution and natural resources extraction that can both raise revenue and further environmental goals. The World Bank and the OECD pro-

* *<Daniel.Slunge@economics.gu.se>*; *<Thomas.Sterner@economics.gu.se>*. Financial support from the Swedish International Development cooperation Agency (Sida) for writing this paper is greatly acknowledged. We would also like to thank our colleagues in Ethiopia, Kenya, Tanzania, Uganda and South Africa for assisting us with background material for the paper. The responsibility for the paper rests with the authors alone.

93

mote EFR as an approach that in addition to raising revenue and generating environmental benefits also can support poverty reduction efforts (World Bank, 2005*a*; OECD, 2005).

The basic idea is that EFR alters the incentive structure and encourages a more sustainable use of natural resources and reduced pollution. Since in low-income countries the poor tend to be relatively more dependent on the natural environment for their welfare an EFR may indirectly contribute to poverty alleviation through improving the environment. In addition, the revenues raised by EFR could be used to finance poverty reduction programmes.

While attractive in theory, few developing countries have undertaken substantial environmental fiscal reforms. In practice EFR is frequently delayed and constrained by political and institutional factors, including a lack of an effective legal, regulatory, and administrative framework and resistance from strong vested interests. Nevertheless, several countries do use a range of different taxes and charges related to environment and natural resources. These instruments have often been introduced for fiscal rather than environmental reasons and in a piecemeal rather than a systematic approach. There is currently a lack of research on both the types of instruments that are used in developing countries and the effects of these instruments on the environment, revenue generation and poverty reduction.

In order to arrive at optimal (or even acceptable) levels of environmental taxation context specific information is needed both on the environmental externality that needs to be addressed and on the distributive effects of different policy instruments. This paper contributes to this generation of knowledge through reviewing the current use of instruments for environmental fiscal reform in selected East and Southern African countries and through analyzing the effects on income distribution from fuel taxes which in terms of revenue generation are by far the most important EFR instrument.

The paper continues as follows. In section 2, the theoretical arguments for introducing taxes on environmental and fiscal grounds as well as the potential trade-off between environmental and fiscal objectives are analyzed. Section 3 provides an overview of the use of environment and natural resource-related taxes and fees in East and Southern African countries. Since data for this type of analysis is not ready available, the section builds largely on a compilation of data from government reports and other sources of so called grey literature. Building on detailed case studies of fuel consumption of different income groups in Ethiopia, South Africa and Kenya, section 4 analyses the distributional effects of fuel taxes. Section 5 concludes.

2. - Environmental and Fiscal Reasons for Environmental Taxation

Price based instruments such as taxes and charges are central in environmental fiscal reform due to their potential to both address environmental pollution and raise revenue. The theoretical argument for environmental taxation is well established in the literature (*e.g.* Baumol and Oates, 1975). When property rights are not fully assigned – that is when there are some "environmental" resources that have no owner – like the air or water – then there is no economic agent who makes sure to economize on these resources. In such cases economic activities – such as production of some sort – which bear its costs in terms of salaries, raw materials and so forth, do not pay for any degradation of these natural resources. However, this degradation may imply very real and economically relevant costs for society, for instance in terms of increased sickness or mortality. The existence of such externalities implies a market failure and economic welfare can potentially be enhanced by environmental policies such as regulation or taxation – assuming that these instruments do not impose more costs than the original environmental damages. When it comes to natural resources (including some that are purely environmental, ecosystem services), there may also be dynamic effects such as stock-dependent growth that affects optimal resource use over time. For instance a healthy ecosystem, a large stock of fish or other animals will produce more growth, and maybe more ecosystem services than a heavily depleted stock. In order to avoid an exaggerated preference for current consumption at the expense of future consumers, a scarcity rent may be motivated to raise the price and to protect the stock for a series of motives: future use, existence values and "breeding" effect of producing more services in the future. In both these cases (a Pigouvian tax and a resource rent), the performance of the economy is thus enhanced by the tax – quite independently of the fact that revenue may be raised that could also be beneficial.

In practice, taxes and charges only make up a subset of a broader menu of environmental policy instruments. In choosing a policy instrument we need to consider not only the environmental problem at hand but also the economic and social context and ask ourselves questions such as: is this a market with competition? Is it an area of rapid technological progress? What are the informational requirements? How will the costs of control and enforcement depend on the choice of policy instrument? We will not go into these issues here but the matrix below provides a simple overview of the many different types of policy instrument available.

TABLE 1

ENVIRONMENTAL POLICY INSTRUMENTS MENU

Price-Type	Rights	Regulation	Info/Legal
Taxes	Property rights	Technological standard	Public participation
Subsidy	Tradable permits	Performance standard	Information disclosure
Charge, Fee/Tariff	Tradable quotas	Ban	Voluntary agreement
Deposit-refund	Certificate	Permit	Liability
Refunded charge	Common property	Zoning	

Table 1 is an attempt at illustrating some of the many dimensions of policy choice. It is not a very satisfactory table since the different policy instruments are not easy to categorize neatly. The table has four columns (– but the rows do not have any particular meaning). Each column represents in some sense a type of policy instrument – but even at this grand level of generality we immediately have problems. We have avoided the term "economic instruments" since economic theory will sometimes lead to the recommendation of for instance zoning or bans or standards – and that is not what people expect when they hear the word "economic" instrument. We have to some extent replaced this with the words like "price-type". Taxes, subsidies and charges are examples of instruments that directly act like a price and in this sense they are "economic" or "market-analogous". But so are tradable permits. What is truly definitive for permits is that they are property and we therefore categorize them in a column for "rights". The column regulation should be fairly self explanatory and contains in fact much of what is the bread and butter of ordinary environmental management by Environmental Protection Agencies around the world. The final column we have referred to as legal or informational includes both labelling and voluntary agreements, liability rules and so forth. Ideally a perfect table should be like the periodic table in chemistry where every element has not only a column but also a row. By that standard we have failed so far to categorize policy instruments, but we hope the discussion somewhat clarifies the choice all the same.

The choice between these instruments depends on a large number of criteria such as their efficiency, flexibility, and the distribution of costs they imply which decides their political feasibility. In order to effectively address environmental problems it is common that a combination of several different environmental policy instruments needs to be used (Sterner, 2003).

There are also fiscal arguments for taxing environmental burdens and exploitation of natural resources. Many low-income countries have recurrent budget deficits and are heavily dependent on development aid not just for the whole economy but specifically to finance the state budget. The fact that a large part of the state budget in many developing countries comes from aid has been criticized from various viewpoints. One of these is that it reduces the need for the government to be accountable to the people and instead creates a false accountability towards donors and creditors. Consequently the need to increase state revenues has been the focus of many reform efforts supported by international financial institutions. Due to the structure of their economies and the limited state capacity, low-income countries face particular challenges in terms of revenue generation. Low-income countries tend to have a large informal economy and a relatively large share of agriculture in total output. This makes income taxes play a relatively less important role for revenue generation in comparison with high income countries. Instead indirect taxes, such as foreign trade taxes, excise taxes on e.g. alcohol, tobacco and fuel and value-added taxes play a relatively more important role. (Di John, 2006; Gupta and Tareq, 2008; Gemmel and Morrissey, 2005). The limited capacity of the state in many low-income countries to collect taxes also tends to favour revenue generation through excise taxes and other indirect taxes which is administratively easier than the collection of income taxes. The number of actual physical tax payers is very limited in most African countries.

However, the idea that EFR would not only solve an environmental problem but at the same time raise revenue in a way that implies little or no excess burden to the economy (the Double-Dividend of environmental taxation) has resulted in considerable controversy. The optimal Pigouvian tax is the tax level at which the difference between consumers and producers *surplus* on the one hand and environmental costs on the other is minimized. The tax-receipts are treated as a purely incidental transfer. If instead these are assigned a value (due to the avoided excess burden that would be incurred through some other form of tax collection) then the optimum tax level would be even higher. It has also been shown by Sandmo (1974) that if society puts a value on tax revenue then the optimal tax on a polluting good may be decomposed into two components; one reflecting the Pigouvian level and then a further component related to the State's need for tax revenue which is basically the taxation to which all goods in the economy (polluting or not) are subject. In everyday terminology this amounts to adding on (for instance) value-added taxes to environmental or natural resource taxes. In the early days of green tax reform, some enthusiasts wanted to go one step further and claim that green tax reform can solve all kinds of problems of improving the tax system to reduce tax evasion, improve incentives for work and thus reduce unemployment and budget deficits in addition to addressing environmental problems. As shown in a series of articles by Bovenberg and others (see Parry, 1998; Parry *et al.*, 1999 or Bovenberg *et al.*, 2002), this will generally be promising too much.

The double-dividend argument is, in its most extreme and simplest form, wrong in the following sense, (Bovenberg and Goulder, 1996): if the environmental benefit (the first "dividend") is disregarded, then an optimal tax structure can hardly be improved by lowering one of the tax rates — say, the rate on wages — and levying a supplemental tax (*i.e.*, beyond what would be motivated by externalities or other market failures) that will give the same additional revenue on an environmental bad. First, it is an illusion that the environmental tax is not a tax on labour; it is indeed a tax on wages because people use their wages to pay whatever good generates the environmental disturbance (such as gasoline), and so, disregarding the externality, it is just an inefficient tax on labour. The inefficiency that arises from lost consumer *surplus* due to a non-optimal distortion of the consumption basket makes the tax effectively bigger in welfare terms than the labour tax it was supposed to replace. Thus, there is a direct welfare loss rather than a gain to this tax swap. To compound this injury, the decrease in effective real salary creates secondary effects, often assumed to reduce labour supply. Another way of looking at this effect is that the environmental "benefit" (as opposed to the loss in private consumer *surplus*) is a kind of public good, and thus, considering the cost of public funds, its provision would be lower in a second-best optimum than the first-best reasoning would lead us to believe.

It is thus not correct to say that an environmental tax would be "a good thing," even if there were no environmental problem. This response has sometimes been used as a last line of defence for environmental tax proposals in which it is scientifically difficult to measure exactly the value of the ecological or health benefits. Interactions between the environmental tax and the pre-existing taxes may be counter-intuitive. Thus, the advantages of the dividends from the environmental tax are not greater in economies with high pre-existing levels of (marginal) labour tax; rather, the high degree of distortion results in greater costs to public funds and bigger deadweight losses.

However, this whole attack on the double-dividend neglects the fact that the environmental externality still has to be addressed. If it is addressed through regulations then even these environmental regulations will have tax interaction effects: if labour taxes are high, then environmental regulations distorting consumption choices will lead to a large welfare loss and a significant reduction in labour supply. Taking this effect into account changes the picture somewhat. The relevant comparison is not just between an environmental tax and a general tax, because a general tax leaves the environmental problem unresolved; instead an environmental tax should be compared with a combination of general tax and an environmental regulation. In this comparison, the environmental tax proceeds are indeed a positive rather than a negative factor.

An optimal environmental tax operates through five effects (Goulder *et* al., 1999):

– the abatement effect, which is fairly self-evident;

– the input substitution effect, which refers to substitution among inputs (in a simple model with few inputs, this effect may be part of the "abatement effect");

– the output substitution effect (products which embody a lot of pollution like aluminium which requires a lot of energy to produce, will become more expensive and be used more sparingly);

– a tax interaction effect, which is related to the loss of consumer *surplus* and real wage that are due to distortion of the preferred consumption (or input) basket. This reduces labour supply and tax incomes, leading to further losses; and

– a revenue-recycling effect, (if an instrument just happens to collect tax revenue then this must be an advantage).

It is this last effect that most clearly corresponds to the positive effect of the environmental tax which is what many people mean when they refer to a double dividend.

Trade-offs between revenue generation and environmental objectives

There are in some cases obvious trade-offs between the objective to generate revenue and address environmental problems with environmental taxes. When abatement is easy, the pollutant emissions may go to zero and thus the tax base would be gone. If the prospect of revenues from such a tax had been used to lower other taxes, there would be a shortfall for the treasury. Because changing the tax system is disruptive and expensive in itself, it is for this reason probably not optimal to base a whole environmental tax reform on environmental problems that are very easily amenable to abatement. Administrative charges or simple information and persuasion (because abatement costs are assumed to be small) may be a more adequate instrument. This case is illustrated in Graph 1B.

GRAPH 1

DEMAND ELASTICITY AND TAX EFFECTS ON EMISSIONS

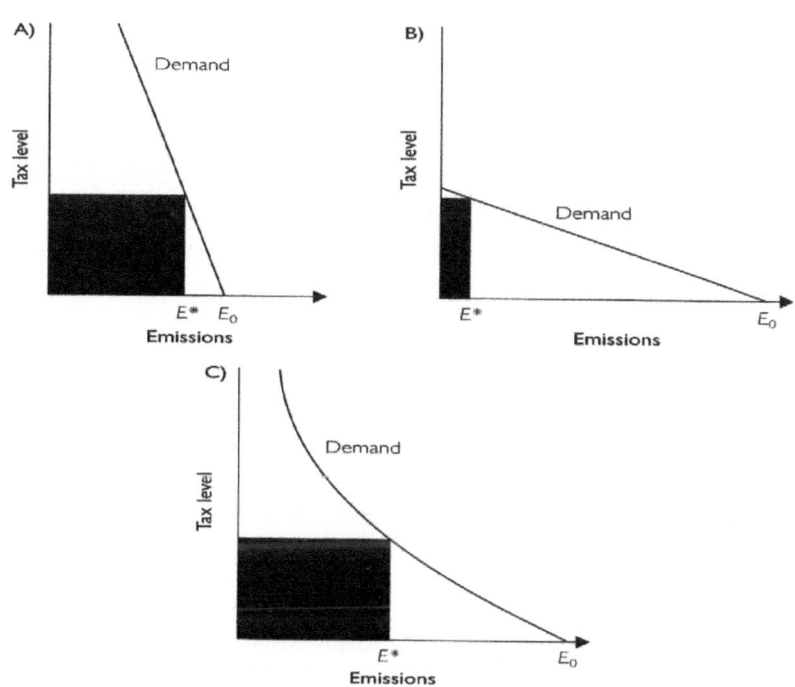

On the other hand, when demand for the polluting good is inelastic (Graph 1A), the tax base is potentially stable but in these cases there is little reduction in pollution. The reason for this is that abatement is difficult, and an environmental tax might appear to be a failed policy. However, because the demand curve is steep, any other policy would also encounter difficulties in achieving significant abatement. At least the tax provides a continuous incentive for innovators as well as for the output substitution and revenue-recycling effects.

Finally, some pollutants fall into the intermediate category where the tax instrument performs well (Graph 1C): first, it provides incentives for abatement through demand reduction; then (as demand becomes more inelastic), it does provide a stable tax base for EFR. In fact this category is very important since it is likely to be a good model for fuel demand.

The energy sector accounts for something of the order of 7 percent of GDP worldwide and is mainly powered by oil and coal that must be phased out almost completely during this century according to our current understanding of the policies needed to address climate change (World Bank, 2010). This is, as pointed out by the Stern Review (Stern, 2006), the largest externality ever encountered. It will be costly to deal with and this is an argument for using the most efficient market based instruments. If a tax is used, enormous revenues will be generated that actually make EFR a very important and interesting concept. We will therefore focus specially on the fossil fuels in one section of this paper.

3. - Environmental Fiscal Reform Instruments in East and Southern African Countries

The use of environment and natural resources related taxes and fees[1] in East and Southern African countries share many sim-

---

[1] An environmental tax is: «A tax whose tax base is a physical unit (or a proxy of it) that has a proven specific negative impact on the environment» (OECD, 2010). Fees and charges are applied more or less in proportion to services provided. The term levy is also used and often covers both taxes, fees and charges.

ilarities. The following overview includes: *(i)* taxes and fees on products and production inputs, *(ii)* taxes and fees to capture the rent from the extraction of natural resources such as forests, fisheries and minerals; *(iii)* fees to recover costs for provision of environmental services; and *(iv)* taxes on emissions. Rather than giving a complete picture of all instruments used in the region, the overview indicates which instruments are commonly used and their relative importance for revenue generation. Table 2 provides an overview of the instruments used.

Environmental taxes and fees on products and production inputs

Fuel taxes and subsidies

All countries in the region tax petroleum products through excise duties on petroleum products. It is common to have a lower tax rate (or even a subsidy) for kerosene which is widely used for cooking and lightning among poorer households. However, if the price difference between kerosene and diesel becomes too large there is an incentive for fuel shifting. In some countries there are special taxes or fees to finance road maintenance, for example the Kenyan road maintenance levy which finances activities of the Kenya Roads Board. Ethiopia and South Africa charge a special fee on petroleum products which finance a fund used to cater for drastic changes in international oil prices.

Fuel taxes are relatively high in most African countries. With the exception of a few oil exporting countries, fuel taxes in African countries are generally well above the level in the US and for many countries even higher than in some countries in the European Union (GTZ, 2009). In recent years some African countries have began to subsidize fuel consumption in order to cushion the impact from the sharp increases in international oil prices. In, for example, Mozambique the government decided to subsidize petroleum prices after that a 50 percent increase in the fares charged by privately-owned minibuses (due to rising fuel prices) had led to riots in February 2008. The subsidies are estimated to amount to approximately 70 million USD or about 2.5 percent of total government expenditures (Metell, 2010). In Ethiopia, the government

TABLE 2

OVERVIEW OF ENVIRONMENTAL AND NATURAL
RESOURCES TAXES AND FEES USED IN EAST
AND SOUTHERN AFRICAN COUNTRIES

Tax Base	Instruments*
Products and inputs	
Transport fuels	Excise tax Levy to finance road maintenance Levy to finance fund for stabilization of fuel prices Subsidies
Motor vehicles	Taxes related to cylinder capacity Charges on imported second hand vehicles
Plastic bags	Tax Ban
Natural resources	
Mining	Production taxes (royalties) Profit taxes Export taxes Sub-national taxes
Fisheries	Access agreements with foreign fishery fleets Vessel licenses Fishermans license Landing sites user fees User levies By-catch fees
Forestry	Royalty on the volume of logged timber Tax related to the area licensed or logged; the standing volume; allowable cut; the property value; the production of forest products. Export taxes Fees for services or materials provided by forest authorities
User fees	
Electricity provision	General electricity levy Electricity tariffs (unitary or progressive; domestic or institutional consumer) Lifeline tariff (free or reduced cost electricity to low income households) Electricity service charge
Water provision	Fees for extraction of raw water from the water source Fees for water resources development and use of water works Water Research fund levy Water consumption tariffs (unitary or progressive; domestic or institutional consumer) Lifeline tariff (some water provided for free to low-income households)
Waste collection	Fees at municipal level
Emissions	Waste water fee

*The table includes examples of instruments used in different countries. See text below to see which instruments are used in specific countries.

has systematically tried to keep fuel prices down. This has resulted in a situation where some petroleum products like kerosene are sold even below cost. This policy in combination with rising international oil prices has contributed to a serious budget deficit. The state owned Ethiopian Petroleum Enterprise which has a monopoly on the importation of petroleum products was forced to borrow over 500 million USD from local banks in order to secure petroleum imports needed for the fiscal year 2008 (GTZ, 2009). In both Ethiopia and Mozambique fuel prices are set administratively and adjusted regularly. The stabilization fund in Ethiopia intended to cater for sharp changes in international oil prices has been emptied of funds and has not been sufficiently strong in light of the rapidly increasing international oil prices (GTZ, 2009). In Tanzania, Kenya and Uganda petroleum prices are liberalized and have increased in line with international oil prices.

Taxes on motor vehicles

Taxes on motor vehicles are also commonly used in East and Southern African countries. In Tanzania and Ethiopia there are annual fees on motor vehicles related to the cylinder capacity. In Tanzania, Uganda and Kenya there are different types of fees on imported second hand vehicles designed to discourage importation of old and more polluting cars.

Taxes on Plastic bags

Recently several countries have introduced taxes on plastic bags. In South Africa the tax rate is quite low and primarily intended to raise revenue to fund plastic recycling (GoSA, 2009). In Kenya and Tanzania the tax rate is 120 percent of the retail price and intended to discourage the use of plastic bags (Ikiara and Mutua, 2007; Mkenda, 2007). Generally the taxes on plastic bags form part of bigger policy packages. In for example Uganda there is an excise duty on plastic bags of more than 30 microns and a ban on other plastic bags (Speck, 2009).

Taxation of Natural Resources

Natural resource extraction and management is another major area where taxes and fees are used.

Mining

In the mining sector most countries use a mix of taxes on production (royalties), profits and exports to capture the resource rents from mineral extraction. Some countries use additional resource rent taxes to enable the country to capture some of the "windfall" profits stemming from periods of exceptionally high mineral prices. In, for example, Malawi a 10 percent resource rent tax is activated when a mining project exceeds a 20 percent rate of return. At the same time it is common that mining companies benefit from tax-exemptions such as import taxes on machinery or other key inputs to production or from higher and faster tax deductible capital allowances which delay the time at which taxes on mining profits begin to be paid. There is a trend towards reduced taxes on production and an increased reliance on taxation of profits (World Bank, 2009).

Recent studies indicate that royalties are generally between 3-5 percent of the value of gross sales for metallic and energy minerals and higher for precious minerals. In for example Tanzania, gold exports, which have increased substantially since the late 1990s, are subject to royalties of 3 percent and in Ethiopia the highest royalty, 5 percent, applies to the extraction of precious metals. South Africa has mainly relied on taxation of profits through corporate income taxes, but a system with royalties was designed to become operational in 2009. (Lambrechts, 2009; Tadesse *et* al., 2009)

Many mining contracts are subject to special agreements between governments and international mining corporations, in which the stated tax rates can be adjusted. The secrecy surrounding many mining contracts makes comparative analyses of tax levels and revenues difficult. Moreover, a large share of the artisanal mining is unregistered and not taxed at all. In for example, Tanzania it is estimated that about 90 percent of more than

500,000 artisanal miners are unlicensed (Ruitenbeek and Cartier, 2007).

Fisheries

Fisheries constitute an important source for livelihoods, export earnings and employment in many coastal African countries. Due to the open access characteristics of the resource and poor management capacity fish stocks are often over-harvested in the region. In combination with other policy instruments, fiscal instruments can play an important part in the management of fisheries. Common instruments used in East and Southern Africa include access agreements with foreign fishing fleets in order to regulate offshore commercial fisheries, export taxes and fees for licenses to operate fishing vessels or to fish.

In Namibia a new regime for fisheries management was introduced following independence in 1990. Through introducing several different policy instruments the new fisheries policy managed to put an end to a situation of open access and uncontrolled fishing on a massive scale. Fishing rights were granted for a period between 7-20 years, fishing vessels licenses were introduced and a total allowable catch (TAC) was set for eight out of 20 commercial fish-species. The TAC was then distributed among the fishing rights holders. Fees on by-catches were also introduced. The instruments introduced in Namibia have also generated substantial government revenue, approximately 16 million USD in 2005 (Nichols, 2003; FAO, 2010).

In Uganda an export industry based on fisheries in Lake Victoria has expanded rapidly and several economic instruments are applied. A Sustainable Fisheries User Levy was introduced in year 2005 consisting of a national export tax at 2 percent of the value of the exported fish, local government level fees for fishing vessel permits and fisherman licences and local level (Beach Management Unit) fish movement permits and landing site user fees. The Fisheries User Levy generated about 2.5 million USD in 2005 out of which about 2 million USD was used for covering costs associated with fisheries management such as fisheries research, monitoring, control and surveillance and market development. The remaining 0.5 million USD of the revenue generated was allocated to the general budget (Ruhweza, 2007).

In Tanzania similar instruments are used, but in 2003 the basis for the export tax changed from 6 percent of the value of exported fish to a specified sum per kilo for different fish species. There is also a fish levy which essentially is a 5 percent sales tax (FAO, 2004).

Illegal, unreported and unregulated fishing represents a major loss of revenue. It also contributes to loss of marine biodiversity and other negative environmental effects, including most importantly a threat to the long run viability of the fisheries (FAO, 2002).

Forests

A mix of instruments is commonly applied by governments to capture the economic rents from the forest sector. The most common type of tax is based on the volume of logged timber. Other types of taxes are related to the area licensed or logged, the standing volume, the allowable cut, the property value, the production of forest products or exports. Fees for services or materials provided by forest authorities are also common (World Bank, 2008).

There are strong indications that revenues collected from the forest sector in many countries are far below the optimum level. Low forest revenues are explained by charges set at too low levels, low collection rates, poor wood measurement, evasion of charges and illegal logging (Gray, 2002). In Tanzania for example it was found that around 58 million USD was lost annually due to under-collection of natural forest product royalties (Milledge *et al.*, 2007). This was mainly explained by widespread illegal activities and evasion of charges. More recently royalties have been increasing in the Tanzanian forest sector. Royalties from timber sales represent 85 percent of income from the sector (amounting to approximately 3.6 million USD). Registration fees for dealers of forest products were the second largest source of revenue. The forest sector in Tanzania is assessed to contribute with about 2 to 3 percent to GDP and 10 to 15 percent to export earnings (World Bank 2008).

Some countries have managed to improve revenue generation from the forest sector substantially. In Cameroon for example, fiscal revenues from forestry grew from USD 3 million to USD 30 million from 1995 to 2001, and provided around 25 percent of national tax revenue. Furthermore, local community returns grew from negligible amounts in 1995 to more than USD 8 million in 2002 (OECD, 2008).

User Fees

In all the reviewed countries different fees, often in the form of tariff based systems, are used with the primary intention of recovering costs for the provision of electricity and water. Waste collection fees are also increasingly being introduced. Tariff structures are often complex and not transparent which makes a comparison between different countries difficult. In addition, since water metering is often limited to urban areas the potential for EFR in the sector is severely restricted. Some examples from the user fees related to the provision of electricity and water are given in the subsequent paragraphs.

Electricity

In South Africa there is a general electricity levy applied at the generation stage and the revenues are earmarked to fund the National Electricity Regulator. Electricity distribution is undertaken by Eskom (the public energy supplying company) and 187 different municipalities. There are differences in the applied tariff structures, but generally, they include a lifeline tariff providing free or reduced cost electricity to low income households (GoSA, 2009). In Ethiopia and Kenya tariffs levied on the consumption of electricity are based on an increasing block tariff scheme (Tadesse *et al.*, 2009). In Uganda there is a flat tariff structure for domestic consumers and electricity consumption has been subsidized during recent years. The subsidy is estimated to have been around 5 percent of total government revenues in the fiscal year 2006/07 (Speck, 2009).

Water

Several countries have undertaken substantial water sector reforms with the objectives of improving water supply and water resources management. Reform efforts have typically included the introduction or revision of fees for water extraction and tariffs for water consumption and waste water, in order to ensure financial sustainability. In Kenya for example water use fees were introduced and tariff structures revised which have resulted in increased collection of revenues (Olum, 2009). In Ethiopia households (in urban areas) are charged with progressive (increasing) tariffs based on the volume of water consumed (Tadesse, 2009). In Uganda households and institutions are charged unitary (flat) tariffs based on the volume of water consumed and sewage produced. A tariff indexation policy was introduced in Uganda in 2002 in order to insulate the tariff against inflation, foreign exchange costs and other external factors, and to ensure that tariffs are sufficient for cost recovery. In, for example, the fiscal year 2008/2009 this indexation resulted in a nominal increase in the average water tariff of more than 9 percent (NWSC, 2010). Also in South Africa, the water pricing strategy includes different levels. The first and second levels relate to the use of water from the water source and water supplied in bulk by the water boards. Several different charges apply which are designed to cover the costs for e.g. catchment management strategies, water resource protection and water demand management. The third level refers to water distribution to end users and tariffs vary between different municipalities. Similar to the provision of electricity in South Africa a lifeline tariff generally assure that some water (typically 25 litres per person and day) is provided for free to low-income households (GoSA, 2009).

Emission Taxes

There are few examples of taxes targeting the actual emissions of pollutants. In Uganda there is a waste water fee amounting to 75 percent of the water charge for households and 100 percent for other users (NWSC, 2010). In South Africa a more complex Waste Water Discharge Charge System is being developed.

Effects on the Environment and Revenue Generation from the Applied Instruments

The overview has demonstrated that various environmental taxes and fees are used in East and Southern African countries. Most of the instruments have been introduced with the primary objective of raising revenue, not correcting for environmental externalities. Despite this, effects on environmental pollution and natural resources extraction do exist but these effects are poorly studied. In as much as forest degradation, excessive energy use or overfishing have been somewhat halted, it is a major benefit.

Among the applied instruments there are large differences in terms of revenue generation. In all countries, fuel taxes constitute by far the largest shares of the revenue generated. In South Africa the general fuel levy has accounted for between 65-74 per cent of all revenue generated from environmentally related taxes between 1998 and 2008 (GoSA, 2009). With a few exceptions, fuel tax is also the only environmental tax that represents a significant share of total tax revenue. In South Africa fuel taxes accounted for 4 percent of total tax revenues 2008/09 (GoSA, 2009). In Kenya and Tanzania fuel taxes constitute about 5-7 percent of total tax revenues, and in Uganda the corresponding share is as high as around 16 percent (Mkenda, 2007; Ikiara and Mutua, 2007; Speck, 2009). In Mozambique revenues from fuel taxes constituted around 12 percent of total tax revenue in 2003 (GTZ, 2005). The same year taxation of fisheries is estimated to have generated revenues corresponding to about 4 percent of total tax revenue and forestry and mining around 0.5 percent of total tax revenue respectively (World Bank, 2005*b*).

The significance of specific sectors for revenue generation is subject to large variations between different countries. In Guinea Bissau, for example, fisheries generate a substantial part of total government revenue. Fishery agreements with the EU are estimated to have generated as much as 30 percent of total government revenue in 2001 (World Bank, 2005*a*). In most countries revenues generated from fisheries related taxes are much more modest. In Tanzania tax revenue from the fisheries sector is estimated to contribute about one percent of total government revenue (FAO, 2004) and in Kenya permit fees for fishing vessels and a tax on fish exports are estimated to have generated only about 0.05 percent of total tax revenue (Ikiara and Mutua, 2007).

Due to poor tax-collection rates, inadequate monitoring and enforcement of existing regulations and in some cases also too low tax rates, revenues collected from natural resources taxation in many countries are far below the optimum level. In the fishing and forest sectors informal and illegal activities signify large losses of government revenue. In the mining sector tax avoidance and poorly negotiated contracts with mining companies cause large amounts of foregone government revenue. In Malawi tax breaks granted to mining companies are estimated to amount to more than 16 million USD annually. In Tanzania 30 million USD annually is an estimate of foregone revenue due to special tax breaks and low royalties with mining companies (Lambrechts, 2009).

From the above analysis we conclude that there is most likely a significant potential to increase the revenues raised from taxation of natural resources. In Mozambique, for example, a World Bank study estimated that tax revenues from fisheries, forests, mining and agricultural land could potentially be increased more than six fold and by 2015 constitute around 19 percent of total tax revenue compared to 5 percent in 2003 (World Bank, 2005*b*). However, the exact magnitude of the revenue potential from taxation of natural resources is hard to estimate.

In addition to the revenue potential from natural resources taxation, there may also be a potential to increase revenues from the taxation of fossil fuels. As shown above, even a small increase in fuel taxes would yield considerable government revenue. Taking into account that most African countries already have relatively high fuel taxes it may sound unlikely that African governments would increase fuel taxes further. However, since energy demand, including the demand for fossil fuel, is strongly correlated with economic growth, fuel demand is projected to increase significantly during the coming decades. In for example Kenya, the total consumption of petroleum products, which was about 3.8 million metric tonnes year 2008, is projected to triple before 2030 (Mutua *et al.*, 2010). In that perspective increased fuel taxes may be an effective policy instrument to internalize the external costs implied by increasing fuel demand. International climate change negotiations may also result in a development where OECD-countries and large developing economies like China, India and South Africa eventually increase their efforts to curb green house gas emissions. If these countries raise their fuel taxes this could encourage African policy makers to increase government revenue through raising fuel taxes. Although low-emitting African countries would not need to increase fuel taxes to meet international obligations to reduce emissions, a higher international carbon price may provide an attractive opportunity for revenue generation.

However, one major concern is the social implications of reducing fuel subsidies and increasing fuel taxes. It is often claimed that fuel taxes are regressive, but growing evidence about the distributional aspects of fuel taxes is questioning this claim. This is the topic of the remaining part of this paper.

4. - The Distributional Impacts of Fuel Taxes

It is generally recognised that fuel taxes are a good tax base. They are moderately effective – there is an environmental effect, but it is not so big as to threaten the tax base itself, at least not in the short or medium term. Quite a large number of developing countries in fact find that taxing petroleum products is easier than taxing many other sources – including income itself. The main concern often voiced is that fuel taxes might be regressive – that is have negative impact particularly on those with lower income levels. In fact there is often considerable resistance to fuel taxes – as there is to any tax – and the lobbies that oppose the tax always make a point of saying the tax is regressive. During 2009, fuel prices were raised in Ghana for instance and immediately there was an outcry in the national papers, asking: what happened to the governmental promises of supporting the poor? The press took it for granted that a fuel tax was bad for the poor.

On the contrary, one might think that in developing countries, cars and gasoline are luxury goods. However also public transport uses diesel and the poor of course use public transport. Similarly all foodstuffs are transported and increased costs of transport might increase costs of essential food. The net effect is far from obvious and it is therefore essential to study this issue in countries with different characteristics. Still the first order effect is of course dominated by the direct consumption of fossil fuels.

The definitive issue when deciding if a tax will be regressive or progressive is to see if the good being taxed is purchased and consumed primarily by the rich or the poor. If the good is used predominantly by the poor – *i.e.* if they have a higher budget share for the good in question than the rich – then a given tax on the good will tend to hit the poor harder than the rich.

The method used here will thus be to study the share of consumption for the good – in this case fossil fuels – by deciles of the income distribution. Data from Kenya for example show very clearly that the direct private use of fuel is very progressive (see Mutua *et* al., 2010). People who belong to the richest decile spend more than 10 percent of their income on private transport whereas the third richest (decile 8) spends 2 percent and all lower deciles much less. Clearly with that pattern, a tax on fuel is more progressive than almost any other feasible alternative tax – be it on income or on value added.

Public transport in Kenya, however, has quite a different pattern. The highest budget shares are for the middle income classes. Yet if all the fossil fuel use (from private and from public transport is aggregated) the result is still progressive although much more weakly so (see Graph 2).

The situation in Ethiopia is in many respects likely to be similar to that in Kenya. The poor use less motor fuel – not just in absolute terms (which is obvious) but even as a percentage of income. This would mean that a tax would be progressive, not regressive. However the opposite might apply for a fuel such as kerosene because it is mainly used by the poor not the rich (who have electricity and other modern fuels for lighting and cooking). We know that the difference in price between kerosene and mo-

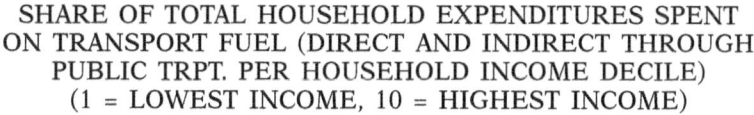

GRAPH 2

SHARE OF TOTAL HOUSEHOLD EXPENDITURES SPENT
ON TRANSPORT FUEL (DIRECT AND INDIRECT THROUGH
PUBLIC TRPT. PER HOUSEHOLD INCOME DECILE)
(1 = LOWEST INCOME, 10 = HIGHEST INCOME)

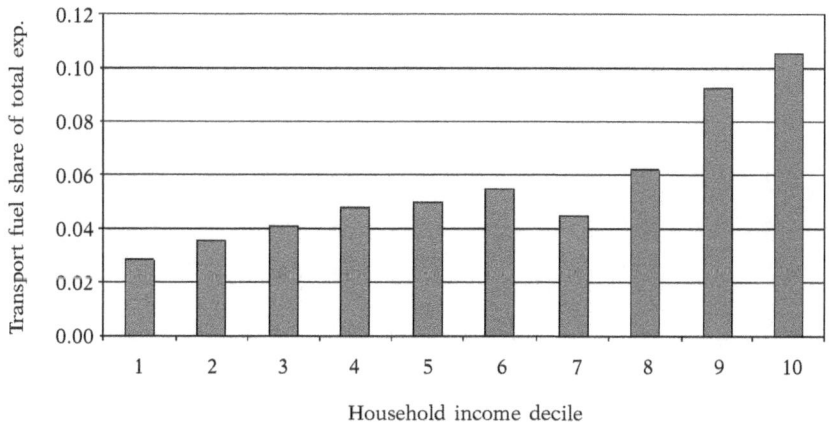

Source: MUTUA J. *et* AL., (2010).

tor fuels cannot be too high – otherwise the incentive for adulterating the motor fuel with cheaper kerosene will be too strong. It might therefore be of interest to study the effect of a tax on both motor fuels and kerosene. Table 3 shows the budget shares for motor fuels and kerosene together for Ethiopia.

It is clear that the budget shares are broadly speaking rising. It is true that the budget share falls marginally from decile 2 to 3 and again from 6 to 7 but the broad picture is still one of increasing budget shares. When we compare the richest we find that they spend more than 2 percent of their expenditures on these fuels while the very poorest spend around 1 percent. A tax on motor fuels would thus be broadly progressive even if it were accompanied by an equal tax on kerosene in the case of Ethiopia.

Finally we turn to the case of South Africa, where average incomes are very much higher. As in the case of Ethiopia this study again uses an expenditure-based measure of lifetime income which

TABLE 3

BUDGET SHARES FOR MOTOR FUELS AND KEROSENE, ETHIOPIA

Income Decile	Total Household Expenditure (USD)	Household Expenditure on Private Transport & Kerosene (USD)	Budget share (percent)
1	2,906.54	33.43	1.15
2	4,245.41	51.92	1.22
3	5,133.94	56.36	1.10
4	5,939.36	76.76	1.29
5	6,768.66	88.34	1.31
6	7,713.65	108.86	1.41
7	8,944.11	116.88	1.31
8	10,636.06	164.29	1.54
9	13,691.47	229.58	1.68
10	29,522.38	611.42	2.07

Source: MEKONNEN A. *et* AL., 2010.

is annual household expenditure. All households were classified by expenditure deciles. (This is because it is argued that households save and borrow and so the current income is a bad indicator of actual lifetime income).

Table 4 shows the *ratio* of fuel expenditure (for own transport services) in total household expenditure. The budget share of fuel generally increases as income increases. The lowest expenditure decile devotes 0.03 percent of their total expenditures to fuel. The highest decile devotes about 3.4 percent of their total expenditures to fuel. The expenditure-based calculations suggest that the distribution of fuel expenditure is generally progressive, with higher income households devoting the highest budget shares to fuel.

Households also make use of fuel indirectly in other transport related activities. This is done through the use of public and

TABLE 4

FUEL EXPENDITURE/TOTAL EXPENDITURE,
BY EXPENDITURE DECILE, SOUTH AFRICA 2000

Expenditure Deciles	Fuel exp/Total expenditure (percent)
1	0.03
2	0.03
3	0.05
4	0.11
5	0.27
6	0.50
7	0.74
8	1.30
9	2.74
10	3.39

Source: CHITIGA M. - MABUGU R. - ZIRAMBA E. (2010).

hired transport. Such transport takes various forms, which include the use of buses, train, rented vehicles, taxis and furniture removal for the transportation of goods. Even counting all indirect uses does not make a fuel tax regressive; see Chitiga, Mabugu and Ziramba (2010).

In summary we find that transport fuel taxes in the regional context of low or middle income countries in Africa appear to be progressive and not regressive as often claimed. If indirect use through public transport and food transports is taken into account, this weakens but does not reverse the result that the fuel taxes are progressive. Similarly if also fuels such as kerosene are simultaneously taxed (to avoid having unsustainable differences in consumer prices that might lead to fuel adulteration), the progressive result is again weakened but – at least in the case of Ethiopia – nor removed or reversed.

There is thus no fairness reason for not taxing fossil fuels. However one should remember that real policy makers may not only be driven by altruistic concern for the poorest; they may be driven by fear of the rich and influential. In this case fuel taxes are problematic – but for the opposite reason: they are good for the poor and bad for the urban middle and upper classes who have real power and who want to protect their private motoring interests.

5. - Conclusions

This paper has reviewed some of the experiences in a number of East and Southern African countries suggesting that there would indeed be reason to explore more the options presented by environmental fiscal reform. Most of these countries share a number of characteristics: they are heavily dependent on local natural resources that are not afforded sufficient protection, since property rights are often incomplete. This applies even to land and all the more so to water, biological, geological and ecosystem resources. To protect against rampant externalities as well as to economise on resource use in the long run, some mechanism for allocating rights and for rent collection is needed. Environmental and natural resource taxation is one of the instruments that can achieve this.

At the same time, these same countries suffer from another significant problem related to their public budgets. They essentially lack well developed tax bases and are unable to raise sufficient taxes domestically to provide even the bare minimum of public goods needed for economic growth and development let alone for welfare. The result of this is lost opportunities when it comes to economic development and in many cases an unhealthy dependence on foreign aid.

Environmental fiscal reform has the potential to be one component that contributes to solving both the problems mentioned above. There has been a fear that there would not be sufficient environmental or resource tax bases and there has been another fear that environmental or resource taxes might hurt the poor. Naturally each country and sector needs to be studied carefully, on a case by case basis. Our brief overview shows that while some use is already made of resource taxes, there are probably considerable unexploited opportunities for further environmental tax reform that could be beneficial for the environment, for fiscal stability and growth and that would not hurt the overall interests of the poor. Naturally, however, one always has to be fairly careful since there may be powerful vested interests who try to steer reform processes to their own advantage.

BIBLIOGRAPHY

BAUMOL W.J. - OATES W.E., *The Theory of Environmental Policy*, Cambridge, Cambridge University Press, 1975.

BOVENBERG A.L. - GOULDER L.H., «Optimal Environmental Taxation in the Presence of Other Taxes: General-Equilibrium Analyses», *American Economic Review*, no. 86 (4), 1996, pages 985-1000.

— - —, — - —, «Environmental Taxation and Regulation», in AUERBACH A. - FELDSTEIN M. (eds.), *Handbook of Public Economics*, New York, North Holland, 2002.

CHITIGA M.R. - MABUGU R. - ZIRAMBA E., «An Analysis of the Efficacy of Fuel Taxation for Pollution Control in South Africa», in STERNER T. (ed.), *Do Gasoline Taxes Hurt the Poor, Progressivity of Fuel Taxes*, Washington DC, Forthcoming RFF Press, 2010.

DI JOHN J., «The Political Economy of Taxation and Tax Reform in Developing Countries», *UNU-WIDER Research Paper*, no. 2006/74, 2006.

FAO, *Stopping Illegal, Unreported and Unregulated Fishing*, Rome, 2002.

— - —, «Fiscal Arrangements in the Tanzanian Fisheries Sector», *FAO Fisheries Circular*, no. 1000, Rome, 2004.

— - —, «Namibia - Fishery and Aquaculture Country Profile», Fisheries and Aquaculture Department, *http://www.fao.org/fishery/countrysector/FI-CP_NA/en*, 2010.

GEMMEL N. - MORRISSEY O., «Distribution and Poverty Impacts of Tax Structure Reform in Developing Countries: How Little We Know», *Development Policy Review*, no. 23(2), 2005, pages 131-144.

GOULDER L.H. - PARRY I. - WILLIAMS R. - BURTRAW D., «The Cost-Effectiveness of Alternative Instruments for Environmental Protection in a Second-Best Setting», *Journal of Public Economics*, no. 72 (3), 1999, pages 329-360.

GOVERNMENT OF SOUTH AFRICA (GoSA), «A Framework for Considering Market-Based Instruments to Support Environmental Fiscal Reform in South Africa», *National Treasury, Tax Policy Chief Directorate, Draft Policy Paper*, 2009.

GRAY J., «Forest Concession Policies and Revenue Systems, Country Experience and Policy Changes for Sustainable Tropical Forestry», *World Bank Technical Paper*, no. 522, 2002.

GTZ, *International Fuel Prices 2005*, Eschborn, 2005.

— - —, *International Fuel Prices 2009*, Bonn, 2009.

GUPTA S. - TAREQ S., «Mobilizing Revenue Strengthening Domestic Revenue Bases is Key to Creating Fiscal Space for Africa's Developmental Needs», *Finance and Development*, 2008, pages 44-47.

IKIARA M. - MUTUA J., «Potential for Revenues via Environmental Fiscal Reforms in Kenya: Emerging practice and some suggestions for EFR-reviews in Low Income countries», *Background Paper* for Conference on Environmental Fiscal Reform in Munich, 2007.

LAMBRECHTS K., *Breaking the Curse: How Transparent Taxation and Fair Taxes can Turn Africa's Mineral Wealth into Development*, Johannesburg, Open Society Institute of Southern Africa, 2009.

MEKONNEN A.- DERIBE R. - GEBREMEDHIN L., «Distributional Consequences of Transport Fuel Taxes in Ethiopia», in STERNER T. (ed.), *Do Gasoline Taxes Hurt the Poor*, Washington DC, forthcoming RFF Press, 2010.

METELL K., *Personal Communication with Sida Economist in Maputo*, 2010.

MILLEDGE S. - GELVAS I. - AHRENDS A., *Forestry, Governance and National Development: Lessons Learned from a Logging Boom in Southern Tanzania*, TRAFFIC East/Southern Africa, 2007.

MKENDA A., «The Scope of Environmental Tax Reform in Tanzania», *Background Paper* for Conference on Environmental Fiscal Reform in Munich, 2007.

MUTUA J. - BÖRJESSON M. - STERNER T., « Distributional Effects of Transport Fuel Taxes in Kenya», in STERNER T. (ed.), *Do Gasoline Taxes Hurt the Poor*, Washington DC, forthcoming RFF Press, 2010.

NATIONAL WATER AND SEWAGE CORPORATION UGANDA (NWSC), *http://www.nwsc.co.ug/index10.php*, 2010.

NICHOLS P., «A Development Country Puts a Halt to Foreign Overfishing», *Economic Perspectives*, no. 8(1), Electronic Journal of the US Department of State, 2003.

OECD, «Glossary of Statistical Terms», *http://stats.oecd.org/glossary/search.asp*, Paris, 2010.

— - —, *African Economic Outlook 2008*, Paris, 2008.

— - —, *Environmental Fiscal Reform for Poverty Reduction*, Paris, 2005.

OLUM J.P., *Water Resources Issues and Interventions in Kenya*, Water Resources Management Authority, Nairobi, ITC Faculty of Geo-Information Science and Earth Observation of the University of Twente, 2009.

PARRY I.W.H., «The Double Dividend: When You Get It and When You Don't», *National Tax Association Proceedings*, 1998, pages 46-51.

PARRY I.W.H. - ROBERTON C.W. - GOULDER L.H., «When Can Carbon Abatement Policies Increase Welfare? The Fundamental Role of Distorted Factor Markets», *Environmental Economics and Management*, no. 37(1), 1999, pages 52-84.

RUHWEZA A., «Application of Environmental Fiscal Reforms and Other Market-Based Instruments for Environmental Management in Uganda: Progress, Challenges and Future Prospects», in GTZ (ed.), *Environmental Fiscal Reform in Developing, Emerging and Transition Economies: Progress & Prospects*, Documentation of the 2007 Special Workshop hosted by the Federal Ministry for Economic Cooperation and Development (BMZ) and the Deutsche Gesellschaft für Technische Zusammenarbeit (GTZ) GmbH, 2007.

RUITENBEEK J. - CARTIER C., *Putting Tanzania's Hidden Economy to Work: Reform, Management and Protection of its Natural Resource Sector*, Washington DC, World Bank, 2007.

SANDMO A., «A Note on the Structure of Optimal Taxation», *American Economic Review*, no. 64, 1974, pages 701-706.

SPECK S., «Options for Promoting Environmental Fiscal Reform in EC Development Cooperation», *Draft Country Report Uganda*, 2009.

STERN N., «The Economics of Climate Change», *The Stern Review*, Cambridge, Cambridge University Press, 2006.

STERNER T., *Policy Instruments for Environmental and Natural Resource Management*, Washington DC, RFF Press, 2003.

TADESSE H. - MEKONNEN A. - STERNER T., *The Structure of Taxation and the Prospects for Environmental Fiscal Reform in Ethiopia*, mimeo, 2009.

WORLD BANK, *Environmental Fiscal Reform. What should be done and How to achieve it*, Washington DC, 2005a.

— - —, «Mozambique Country Economic Memorandum. Sustaining Growth and Reducing Poverty», *Report*, no. 32615-MZ, 2005b.

WORLD BANK, *Forests Sourcebook - Practical Guidance for Sustaining Forests in Development Cooperation*, Washington DC, 2008.

— - —, «Malawi Mineral Sector Review. Source of Economic Growth and Development», *Report*, no. 50160-MW, 2009.

— - —, «Development and Climate Change», *World Development Report 2010*, Washington DC, 2010.

Environmental Quality and Income Inequality: The Impact of Redistribution on Direct Household Emissions in Italy

Laura Castellucci - Alessio D'Amato - Mariangela Zoli*

University of Rome "Tor Vergata"

This paper investigates the relation between income distribution and direct households' emissions in Italy. Our results seem to confirm some recent articles concerned with income-pollution relationship in other countries. Indeed, our empirical analysis shows that decreasing inequality would lead to higher aggregate emissions, whereas increasing inequality would reduce environmental problems. By going into a deeper inquiry of such results, we identify some weaknesses in the framework proposed by the literature, namely the shape of emission intensities distribution. We show that changes in such distribution might lead to opposite conclusions. [JEL numbers: Q01; Q56; D12]

Keywords: emissions; income inequality, household consumption.

1. - Introduction

The literature on the links between public policy design and environmental quality is now well established, yet some issues, such as income redistribution, lack a full understanding. We focus on the impact of income distribution on environmental degra-

* <*laura.castellucci@uniroma2.it*>; <*damato@economia.uniroma2.it*>; <*zoli@ se-femeq.uniroma2.it*>, Department SEFEMEQ, Faculty of Economics, University of Rome "Tor Vergata". The authors want to thank the referees for extensive comments on an earlier version of the paper. Any remaining errors are our own. Castellucci gratefully acknowledges professor Anil Markandya for having launched the idea of the special issue and gone through the lengthly process of producing it. Special thanks are also due to professor Piga as the managing editor of the Review.

dation. Indeed, recent literature suggests that income distribution can significantly affect environmental pressure. The existence of a potential trade-off between reducing inequality and controlling pollution implies that redistributive policies may have unintended consequences on aggregate emissions. Empirical evidence, however, is mixed. This paper adds to the existing literature by investigating the income-pollution relationship in Italy, with specific reference to how income distribution affects aggregate emissions related to direct household consumption.

Even though the relation between income and environmental quality has been widely explored in the 1990s (Borghesi, 2003), only few works have emphasized the importance of income distribution in explaining environmental outcomes (see, for instance, Boyce, 1994; Torras and Boyce, 1998; Scruggs, 1998; Ravallion *et al.*, 2000; Heerink *et al.*, 2001; Brännlund and Ghalwash, 2008). In a political-economy framework, Boyce (1994) sets forth the hypothesis that the extent of an environmentally degrading activity depends on the balance of power between those who benefit from the activity and those who bear the costs. When the winners are more powerful than the losers, more environmental degradation will occur. Indeed, greater equality of incomes leads to lower levels of environmental degradation. This conclusion has been challenged by other authors. Heerink *et al.* (2001), for instance, finds that higher inequality reduces environmental pollution according to several indicators analyzed on a cross-section of different countries. Their result is based on the argument that an aggregate analysis omitting a measure of income dispersion as explanatory variable will result in biased estimates when the pollution-income relationship is non-linear at the individual level. Specifically, if there is a concave (convex) relation between income and environmental pressure, an income redistribution from the rich to the poor leads to a higher (lower) environmental damage. Then there is no way to determine *a priori* if an income equalization policy is beneficial or not for the environment, since the outcome depends on the shape of the income-pollution relation. To account for the non-linearity bias, in Heerink *et al.* (2001) the Gini coefficient is included in the regression equations which estimate the overall impact of income inequality on the environment. In the same line, Brännlund and Ghalwash (2008) estimate a structural model for consumer demand in order to assess how changes in income distribution affect aggregate emissions through changes in household consumption baskets. On the basis of cross-sectional data for Sweden, they conclude that the pollution-income relationship is strictly concave for all types of pollutants they consider, implying that an income equalization would lead to higher emissions.

In this work we aim at contributing to this recent debate. Specifically, following Brännlund and Ghalwash (2008), we want to test if the same pollution-income relationship exists also in Italy. To this end, we firstly estimate a consumer demand system for Italian households, in order to derive income elasticities at micro level. Then we calculate direct emissions related to household consumption and evaluate emission changes due to an income redistribution. We focus on four different pollutants: carbon dioxide (CO_2), sulphur oxides (SO_x), nitrogen oxides (NO_x) and particulates (PM_{10}), the main polluting agents produced by final household consumption. Our main conclusions support the idea that redistribution reducing (increasing) inequality implies larger (smaller) direct emissions from households. Such conclusion is, however, shown to depend crucially on the assumption about the value of the emission intensity (constant along the income distribution, as in Brännlund and Ghalwash, 2008). We question such assumption and show that completely different results about the environmental impact of income redistributions can be obtained by adopting other, equally possible, emission intensity distributions. In the absence of a detailed dataset necessary to empirically estimate the income-emission intensity relationship, we suggest that the trade-off between environmental and equity goals identified by the literature should be better investigated.

The rest of the paper is organized as follows. In Section 2 we discuss the income-pollution relationship and present the empirical model. Section 3 describes consumption and emission data, whereas Section 4 provides the main results of our empirical analysis. Finally, Section 5 discusses previous results and shows relevant counterexamples, while concluding remarks are given in Section 6.

2. - The Income-Pollution Relationship

Households' consumption implies emissions of several pollutants. Such emissions can be related either "directly" to the consumption of certain goods or "indirectly" through their production. Obviously, any change in consumption patterns due to an income change has an impact on environmental quality. The sign of this impact, however, is rather ambiguous. Rising incomes, for instance, may increase the demand for some polluting goods (such as heating and transport); at the same time, richer households may reduce their demand or rely on modern (and less polluting) appliances. It implies that the relation between income and environmental degradation may be non-linear at the household level. As a result, an income redistribution reducing (or increasing) the degree of inequality in the population may affect aggregate emissions in an ambiguous way.

Following Brännlund and Ghalwash (2008), we assume that emissions produced by a household i are a function of the household income (Y_i). The average level of emissions per household ($\bar{E}$) can then be expressed as:

$$\bar{E} = \frac{1}{n}\sum_{i=1}^{n} f(Y_i) = f(\bar{Y}) + \frac{1}{n}\sum_{i=1}^{n}\left\{ f(Y_i) - f(\bar{Y}) \right\}$$

where $\bar{Y}$ is the average household income. The way in which an income redistribution affects aggregate emissions will depend on the properties of the function f (Heerink *et al.*, 2001). If the function f is strictly convex, reducing inequality will lower average emissions; on the contrary, if the function f is strictly concave, aggregate emissions will be increased as a result of an equalizing redistribution. As shown in Brännlund and Ghalwash (2008), non-linearities in the income-pollution function can be introduced *via* the income-consumption relationship. In Brännlund and Ghalwash (2008), the shape of the income-pollution function depends on the derivative of the income elasticity with respect to income. Accordingly, a consumer demand system is estimated in order to evaluate how changes in income distribution affect the environmental quality through changes in the households' consumption bundles.

Given the Quadratic Almost Ideal Demand System (QUAIDS) specification (Banks et al., 1997), which allows for the presence of non-linearities in the demand model, estimating the household income elasticity requires to estimate the following expenditure share equation system:

$$(1) \qquad s_{ij} = \bar{\alpha}'_j d_i + \beta'_j d_i \ln\left(\frac{Y_i}{P}\right) + \delta'_j d_i \ln\left(\frac{Y_i}{P}\right)^2 + v_{ij} \qquad j = 1,\ldots,$$

where s_{ij} is the budget share for good j and for household i. Household characteristics are summarized by a vector of dummy variables (d_i), while v_{ij} is the residual term. It follows that the income elasticity is given by:

$$(2) \qquad \varepsilon_{ij} = \frac{1}{s_{ij}}\left[\beta'_j d_i + 2\delta'_j d_i \ln\left(\frac{Y_i}{P}\right)\right] + 1 \qquad j = 1,\ldots,k$$

In other terms, the quadratic logarithmic specification implies that the income elasticity depends on the income level, meaning that some goods may be considered as necessities at some income levels and luxuries at others. The household income elasticity estimation allows for estimating the effect of an income change on the demand for the various goods.

Whereas the income-consumption relation is non-linear, emissions are considered as a linear function of household consumption. The change in aggregate emissions due to an income change can then be calculated as:

$$(3) \qquad \frac{\partial E_m}{\partial Y} = \sum_{j=1}^{k} \frac{\partial x_j}{\partial Y}\frac{\partial E_m}{\partial x_j} = \sum_{j=1}^{k} \frac{\partial x_j}{\partial Y}\theta_m$$

where m indicates the type of pollutant and $\frac{\partial E_m}{\partial x_j}$ is the emission intensity for each substance (total emissions per unit of real consumption of each good). Following Brännlund and Ghalwash (2008), we assume $\frac{\partial E_m}{\partial x_j} = \theta_m$ for pollutant m, i.e. emission intensities are constant. In Section 5 we will discuss the implications of such assumption.

3. - Data

To estimate the income-pollution relationship in Italy we need data on both household consumption expenditure and emissions associated to each kind of consumption good.

The expenditure data are taken from the Italian Household Budget Survey (IHBS) for 2005, including a random sample of 24,107 households throughout the country. This survey, which is conducted by the National Institute of Statistics (ISTAT), is one of the most comprehensive sources of microdata on consumption behavior in Italy, yielding detailed information on family expenditures as well as on household socioeconomic and demographic characteristics.

Both non-durable and durable consumption data are provided in the survey. Nevertheless, since data on direct emissions are available only for certain categories of goods, as will be explained later on, we restrict our analysis by considering only expenditure on non-durables. This is coherent with Brännlund and Ghalwash (2008), where it is assumed a two-stage process in the income allocation between durables and non-durables.[1]

For non-durables and services, household expenditures are collected over a one-week period and then expressed on a monthly basis. The collection of information on a seven-day period introduces some room for undetected infrequency of purchases. It means that an observed zero expenditure does not necessarily mean that the household does not make such expenditure, but simply that such good has not been purchased in the considered period.[2]

---

[1] For further details, see BRANNLUND R. and GHALWASH T. (2008).

[2] In order to not introduce distortions in households' behavior, we simply drop households (records) reporting a zero value for some budget share. The final number of observations is then equal to 17,689 households.

Emission data are provided by the 1990-2006 time series of NAMEA (National Accounting Matrix including Environmental Accounts), produced by ISTAT. This database includes the emissions of eighteen air pollutants broken down by economic activity and households' consumption expenditure. Specifically, "direct" emissions (*i.e.* related to direct households' consumption) are divided among three main sources: transport (within which only expenditure on fuels and lubricants for personal transport equipment can be related to emissions)[3], heating (including expenditure on electricity, gas and other fuels)[4] and other expenditures (among which only activities related to varnishing and solvent use produce emissions). "Indirect" emissions (*i.e.* those obtained by using emission data from the production side) cannot be directly related to households consumption expenditures. Accordingly, in this work we focus on direct emissions of carbon dioxide (CO_2), sulphur oxides (SO_x), nitrogen oxides (NO_x) and particulates (PM_{10}), the main polluting agents produced by final consumption. Furthermore, for guaranteeing data compatibility, we consider emissions for 2005.

TABLE 1

BUDGET AND EMISSION SHARES
(*year 2005*)

Budget share	Emission share				Emission intensities			
	CO_2	NO_x	SO_x	PM_{10}	CO_2	NO_x	SO_x	PM_{10}
Transport 14.0	46.7	75.2	6.4	45.8	1.79818	0.00573	0.00003	0.00050
Heating 13.0	52.8	24.8	93.6	54.2	1.87047	0.00174	0.00039	0.00055
Others 73.0	0.5	—	—	—	0.00077	—	—	—
Sum 100.0	100.0	100.0	100.0	100.0				

Note: Percentage of total expenditure and total emissions. Emission intensities are in tons/euro. The contribution of "other expenditures" to direct emissions is negligible for CO_2 and is not reported for other pollutants.

---

[3] COICOP (Classification of Individual Consumption According to Purpose) code 07.2.2.

[4] COICOP code 04.5.

Expenditure shares for each good as well as emission shares and intensities in 2005 are reported in Table 1. Transport and heating represent almost the same proportion of total households' expenditure and contribute for almost half of both total CO_2 and PM_{10} emissions. Transports are the main responsible expenditure for NO_x pollution, while heating produces the largest contribution of SO_x emissions.

4. - Empirical Results

In this Section we empirically analyze the income-pollution relationship in Italy, by showing how an income redistribution affects environmental quality in the considered year. Accordingly, we firstly estimate the household demand model and then use the parameter estimates to calculate income elasticities.

To estimate the demand model, we consider the system defined above (eq. *1*) for the three goods related to direct emissions (transport, heating, other expenditures). Prices are considered as constant in regressions[5] and are included in the intercept term $\tilde{a}$. Given the categories of commodities we are considering, only two typologies of household characteristics, namely household composition and the area of residence, have been deemed relevant in affecting consumer behavior. Accordingly, regressions include three dummy variables for household composition (couple without children, couple with one child, other family types – households with two children or more and single parents), and one geographical dummy variable for households living in the North/Centre of Italy.[6]

Finally, and coherently with recent works[7] (see, for instance, Poterba, 1991 and West, 2004), total expenditure on non durable goods has been used as a proxy of household income.

---

[5] We have adopted national prices, in the absence of detailed information on prices for transport and heating fuels in different Italian regions for 2005.

[6] Our reference category is represented by households including only one person, living in the South.

[7] In these works, total consumption expenditure is considered as a good proxy for permanent or lifetime income.

The demand system has been estimated by using the Seemingly Unrelated Regressions (SURE) model, which allows for disturbances to be correlated across observations. In our case, since explanatory variables are the same in all equations, parameter estimates (coefficients) are identical either by estimating each equation separately with ordinary least squares or estimating all equations simultaneously with SURE. The SURE estimate, however, produces more efficient standard errors, and allows for testing non linearities in the model specification. Such test is crucial in our setting, since the quadratic specification significantly affects the shape of

TABLE 2

PARAMETER ESTIMATES FROM THE DEMAND MODEL IN 2005
(*t-ratio* within parentheses)

	Heating		Transport	
Intercept of the expenditure equation				
Constant	1.24	(5.62)	0.36	(1.48)
Couple	-1.13	(-4.23)	0.02	(0.07)
Ch1	-0.88	(-2.95)	-0.40	(-1.22)
Other_fam type	-0.57	(-2.35)	-0.07	(-0.27)
North_centre	-0.21	(-1.22)	0.49	(2.57)
Linear expenditure coefficients				
Constant	-0.31	(-4.57)	0.02	(0.32)
Couple	0.36	(4.4)	-0.05	(-0.52)
Ch1	0.28	(3.16)	0.08	(0.79)
Other_fam type	0.19	(2.6)	-0.03	(-0.35)
North_centre	0.06	(1.17)	-0.13	(-2.31)
Quadratic expenditure coefficients				
Constant	0.02	(3.98)	-0.01	(-1.25)
Couple	-0.03	(-4.51)	0.006	(0.85)
Ch1	-0.02	(-3.3)	-0.003	(-0.39)
Other_fam type	-0.01	(-2.74)	0.005	(0.88)
North_centre	-0.004	(-1.02)	0.009	(2.07)

Note: Couple, without children; Ch1, couple with 1 child; Other_famtype, couple with more than 1 child, single parent, other family types.

the income-pollution relationship. The likelihood-ratio test applied on our expenditure data suggests that the hypothesis of linearity can be rejected. Estimated parameter values from SURE regressions are provided in Table 2. Results reveal that the family type is statistically significant for heating but not for transport, whilst the geographical area of residence is relevant only for transport.

The parameter values estimated from the demand model are then used to calculate households' income elasticities of each expenditure component (eq. 2). Resulting average elasticities, *i.e.* evaluated at the mean budget share and the mean total expenditure, are provided in Table 3. It follows that both heating and transport are normal goods and can be considered as necessities, since elasticities are respectively roughly equal to one and lower than one.

TABLE 3

ESTIMATED DEMAND ELASTICITIES

	Budget elasticity
Heating	1.0551 (.025)
Transport	0.9448 (.018)

Note: Standard error are in parentheses.

On the basis of estimated income elasticities, we can investigate the relationship between pollution and income by examining how a change in the income distribution may affect aggregate emissions.

We consider the empirical distribution of households' total expenditure on non-durable goods in 2005. Given the observed distribution (Graph 1), it can be assumed that total expenditure follows a lognormal distribution, *i.e.* $\ln y \sim N(m, s)$, where m is the mean and s is the standard deviation.[8] In order to determine the

---

[8] The estimated mean and standard deviation are respectively equal to 6.849 and 0.492. Given this values for the lognormal distribution, the mean ($\bar{y}$) and standard deviation (σ) for y are equal to 1064.86 and 557.67.

impact of an income change on aggregate emissions, we replicate the exercise carried out in Brännlund and Ghalwash (2008) and simulate a change in the expenditure distribution. Specifically, we increase/decrease the value of *s* in the lognormal distribution while adjusting the value of *m* to keep average expenditure unchanged. In this way we simulate a rise/reduction of the overall inequality, compared to the reference case. Results are displayed in Table 4.

GRAPH 1

DISTRIBUTION OF TOTAL EXPENDITURES ON NON-DURABLE GOODS IN 2005

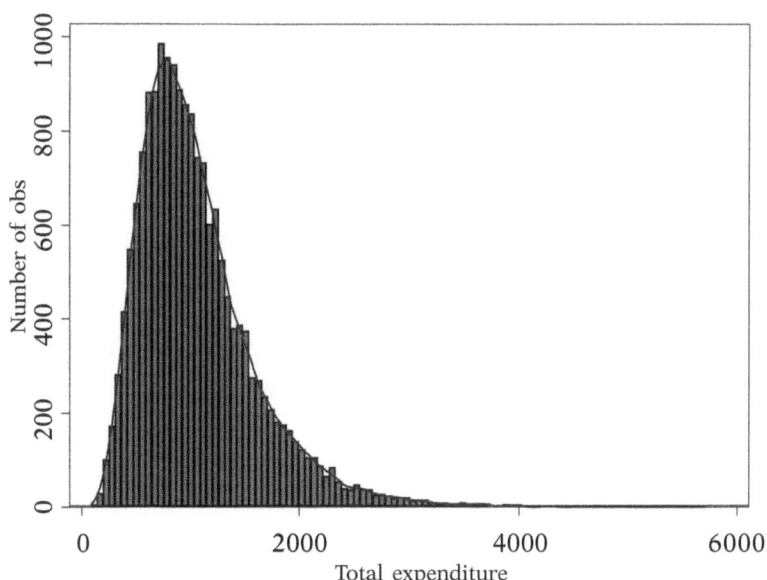

The low variance case corresponds to a standard deviation *s(low)* = 0.5*s*, whereas the high variance case is defined by *s(high)* = 1.5*s* (corresponding standard deviation values for *y* are indicated with σ in Table 4). The change in the degree of inequality is expressed by the coefficient of variation, defined as the *ratio* between the standard deviation and mean expenditure.

TABLE 4

AGGREGATE EMISSIONS IN DIFFERENT
INCOME DISTRIBUTION SCENARIOS

	Low variance	Reference	High variance
Std deviation (s)	0.25	0.49	0.74
Std deviation (σ)	242.7	557.67	1055.13
Coefficient of variation	0.23	0.52	0.99
CO_2	10138.46(+3.66%)	9780.10	9327.27(-4.63%)
NO_x	22.85(+2.21%)	22.36	21.66(-3.11%)
SO_x	0.96(+6.71%)	0.90	0.83(-7.82%)
PM_{10}	2.89(+3.73%)	2.78	2.65(-4.70%)

Note: Emissions are in thousands of tons.

Changes in the income distribution imply changes in households' consumption patterns on the basis of estimated income elasticities (eq. *2*); changes in consumption in turn affect the emissions of each pollutants, according to eq. *3*. As Table 4 reveals, a
reduction in the degree of inequality would lead to an increase in
the level of aggregate emissions for all types of pollutants, whereas a higher inequality would reduce total pollution. The magnitude of these changes is particularly relevant for SO_x, which is
mainly related to fuel consumption for heating.

5. - Discussion

Previous results suggest that in Italy, as well as in other countries (Brännlund and Ghalwash, 2008; Ravallion *et* al., 2000), a
trade-off between reducing income disparities and controlling polluting emissions seems to exist. The policy implication that logi-
cally follows – *i.e.* reducing income inequality is not desirable because it can be detrimental for the environment – clearly contrasts
with distributive goals. It is then crucial to test the robustness of
previous considerations.

In this Section, we focus on relaxing the supposed assumption that the emission of each substance per unit of real consumption is constant along the income distribution. Intuitively, we can presume that the emission intensity associated to households' consumption can vary across different income deciles due to different environmental attitudes and the adoption of less-emission intensive technological devices.

Let us then suppose that emission intensities are a function of income, *i.e.* $\theta_m = \theta_m(Y)$, and modify eq. *(3)* to account for such a change:

(4)
$$\frac{\partial E_m}{\partial Y} = \sum_{j=1}^{k} \frac{\partial x_j}{\partial Y} \theta_m + \sum_{j=1}^{k} x_j \frac{\partial \theta_m}{\partial Y}$$

Comparing *(3)* with *(4)* we can expect that a redistribution reducing inequality might, in principle, lead to a decrease in total emissions. This may happen, for instance, when θ_m increases with income, so that redistributing to the poor implies a decrease in the environmental impact of households' consumption. But several distributions for θ_m can be imagined leading to the same result.

Unfortunately, to the best of our knowledge, available datasets do not allow us to empirically estimate the relationship between income and emission intensities. Accordingly, we cannot assume a specific emission intensity distribution for income levels based on the empirical evidence and we cannot draw any definite conclusion about the potential impact on the environment of income equalizing policies. It can also be expected that emission intensities vary with income in different ways for different consumption goods.

By limiting our attention to Italy, an approximate picture of the relationship between income and emission intensities can be provided by considering the relationship between *per capita* disposable income and emission intensities at the regional level.[9] Graph 2 shows emission levels per unit of heating expenditure

---

[9] Emissions data for all Italian regions are provided by ISTAT only for one year, that is 2005.

produced by Italian regions: on the orizontal axis of each graph the regional *per capita* disposable income is reported, whilst emission intensities are represented on the vertical axis. Graph 3 displays the same scatter plot for regional emission intensities generated by transport expenditure. As it can be seen, patterns are different for the two types of consumption expenditure: except for PM_{10}, a positive linear correlation between emission intensities and income can be identified for all types of pollutants when heating expenditure is taken into account. At the opposite, for transport expenditure, a negative linear correlation emerges for all substances. Coefficients and R-squared of each regression are reported in Table 5.

GRAPH 2

EMISSION INTENSITIES FOR HEATING EXPENDITURE
IN ITALIAN REGIONS (YEAR 2005)

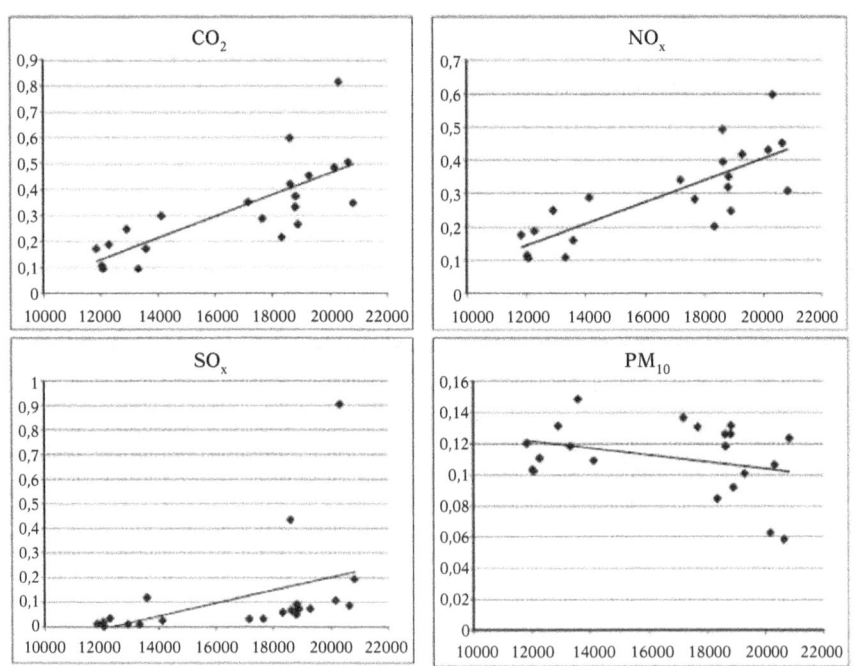

GRAPH 3

EMISSION INTENSITIES FOR TRANSPORT EXPENDITURE
IN ITALIAN REGIONS (YEAR 2005)

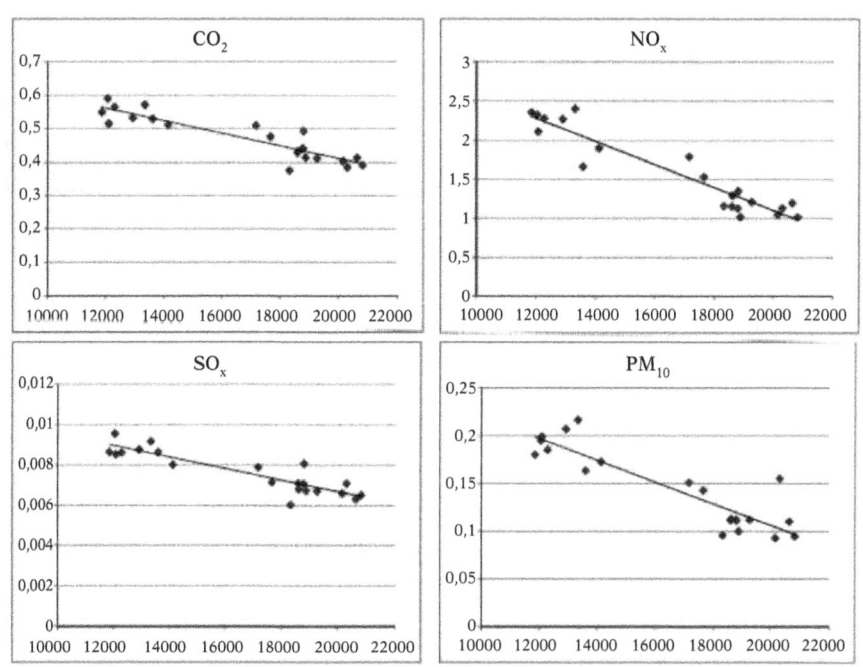

TABLE 5

CORRELATIONS BETWEEN REGIONAL *PER CAPITA*
INCOMES AND EMISSION INTENSITIES (2005)

Pollutant	Coef.	*t* values	*R*-squared
CO_2_heating	4.15E-05	5.09	0.5773
NO_x_heating	3.24E-05	5.65	0.6271
SO_x_heating	3.24E-05	2.07	0.1837
PM_{10}_heating	-2.2E-06	-1.47	0.1023
CO_2_transp	-1.9E-05	-9.48	0.8254
NO_x_transp	-0.00015	-12.59	0.8929
SO_x_transp	-2.8E-07	-8.83	0.8042
PM_{10}_transp	-0.00015	-8.6	0.7957

By taking these patterns as a proxy of the relationship between income and emission intensities at the micro level, it can be imagined that a redistribution reducing inequality may tend to reduce overall emissions related to households' consumption of heating and to increase those associated to transport expenditures. The net effect is obviously ambiguous.[10] When we adopt the parameter values estimated at regional level for simulating the effect of income redistributions on households' total emissions, we find that, at least for SO_x, reducing the degree of inequality would lead to a decrease in aggregate emissions. In this case, the reduction in total emissions related to consumption of heating fuels is only partially offset by increased emissions from transports (see Table 6).

TABLE 6

SO_X EMISSIONS (LINEAR CORRELATION HYPOTHESIS)

SO_x (percentage change)	Low variance	High variance
Heating	-32.97%	13.77%
Transport	1.65%	-3.21%
Total	-19.77%	7.30%

Due to the unavailability of more detailed data we can only simulate the environmental effects of redistributive policies under different assumptions about the income-emission intensity relationship. In Table 7, we report the change in overall CO_2 emissions obtained by altering the observed income distribution, under different scenarios about the emission intensity-income relationship. In the first case θ_{CO2} is assumed to be increasing and convex in income, whilst in the second case emission intensities are assumed to be higher for the lowest and the highest incomes[11].

---

[10] Aggregate emissions are clearly affected also by the way in which income changes affect the demand for the various goods, through income elasticities.

[11] This pattern reflects the hypothesis that lower income households tend to adopt older, high-intensity technologies, whereas richer households tend to use up-to-date technologies which however can be high-intensity as well (as for cars).

In both cases, results are completely reversed compared to those in Table 4, *i.e.* reducing inequality improves direct households' emissions and *vice versa*.

TABLE 7

CO_2 EMISSIONS UNDER DIFFERENT EMISSION
INTENSITY SCENARIOS

CO_2 (percentage change)	Low variance	High variance
Convex function hypothesis	-16.2%	21.5%
Piecewise concave function hypothesis	-31.8%	73.8%

These outcomes then suggest that the existence of a trade-off between environmental and equity goals may be questionable since it strongly depends on the assumptions of the model. They highlight also the need for a more detailed dataset in order to empirically estimate how emission intensities vary over the income distribution. In its absence, any assumption about the shape of the relationship between emission intensities and income appears as arbitrary as the hypothesis that emission intensities are fixed and independent of income.

6. - Concluding Remarks

This paper rests on the recent literature focusing on the relevance of income distribution for correctly interpreting the income-pollution relationship. Specifically, our work is based on the analysis carried out in Brännlund and Ghalwash (2008), where the link between consumption and pollution is empirically assessed for Sweden. By replicating their analysis for Italy we obtain similar results for the income-pollution relation. Specifically, our empirical analysis shows that an income redistribution reducing inequality would lead to higher emissions, whereas an increased inequality scenario would mitigate environmental problems.

As for Sweden, then, our results suggest that a higher inequality scenario can be beneficial for the environment. Nevertheless, such conclusion follows from some intrinsic limitations of the empirical framework proposed by the literature, the most important being the assumption that emission intensities are constant and independent of income. By relaxing this hypothesis, we show that completely different conclusions can be drawn, and reducing inequality can lower aggregate emissions. A full understanding of the link between inequality and households' direct emissions cannot go without an empirical estimation of the income-emission intensity relationship, which in turn requires a specific data set presently not available.

BIBLIOGRAPHY

BANKS J. - BLUNDELL R. - LEWBEL A., «Quadratic Engel Curves and Consumer Demand», *The Review of Economics and Statistics*, Vol. LXXIX, no. 4, 1997, pages 527-539.

BORGHESI S., «Disuguaglianza di reddito, crescita economica e degrado ambientale», *Studi e Note di Economia*, Vol. 1, 2003, pages 131-151.

BOYCE J.K., «Inequality as a Cause of Environmental Degradation», *Ecological Economics*, no. 11, 1994, pages 169-178.

BRANNLUND R. - GHALWASH T., «The Income-Pollution Relationship and the Role of Income Distribution: An Analysis of Swedish Household Data», *Resource and Energy Economics*, Vol. 30, no. 3, 2008, pages 369-387.

HEERINK N. - MULATU A. - BULTE E., «Income Inequality and the Environment: Aggregation Bias in Environmental Kuznets Curves», *Ecological Economics*, no. 38, 2001, pages 359-367.

POTERBA J., «Is the Gasoline Tax Regressive?», in BRADFORD D. (ed.), *Tax Policy and the Economy*, Cambridge, MA, MIT Press, 1991, pages 145-164.

RAVALLION M. - HEIL M. - JALAN J., «Carbon Emissions and Income Inequality», *Oxford Economic Papers*, no. 52, 2000, pages 651-669.

SCRUGGS L.A., «Political and Economic Inequality and the Environment», *Ecological Economics*, no. 26, 1998, pages 259-275.

TORRAS M. - BOYCE J.K., «Income, Inequality, and Pollution: A Reassessment of the Environmental Kuznets Curve», *Ecological Economics*, no. 25, 1998, pages 147-160.

WEST S., «Distributional Effects of Alternative Vehicle Pollution Control Policies», *Journal of Public Economics*, no. 88, 2004, pages 735-757.

III - CARBON TAXATION

Carbon Pricing as an Effective Instrument of Climate Policy: Searching for an Optimal Policy Instrument

Alberto Ansuategi - **Ibon Galarraga**[*]

University of the Basque
Country (UPV-EHU)

Basque Centre for
Climate Change (BC3)

The paper highlights the use of carbon pricing as an effective market tool to control GHG emissions It describes the discussion on the benefits and limitations of both carbon tax and carbon trading. The paper summarises this debate and it argues that, as well as economic effectiveness, many other factors are highly relevant when choosing the right policy instrument. These include political feasibility, impact on competitiveness, institutional requirements, incentives for R&D and many others. Finally, interesting lessons are drawn for policy makers and academics willing to develop such instruments. [JEL: Q54, Q58, H00]

Keywords: carbon taxes; cap and trade.

1. - Introduction

Climate change has attracted a great deal of attention among policy makers in the last decade as the greatest and widest-ranging market failure ever seen (Stern, 2006). The wide range of particularities attached to it (global approach, great uncertainty, the expected impacts being very long term and the specific nature for risk management) pose a great challenge for both economists and

---

[*] *<alberto.ansuategi@ehu.es>* Foundations of Economic Analysis I. Faculty of Economics. *http://www.ehu.es*; *<ibon.galarraga@bc3research.org> www.bc3research.org.*

politicians wishing to design effective instrument mixes to fight said change.

Among the policy options considered, the traditional distinction between command – and – control *versus* market-based instruments has led the way to more detailed discussion regarding the main two types of "Carbon Pricing" schemes: carbon tax and carbon trading. Both instruments have been used in other environmental resource management policies for many years with significant success. The size of the challenge in the field of climate change is significantly larger than any other faced in human history. This paper helps to unravel the ensuing far-ranging debate.

The following section deals with the rationale for pricing instruments while sections 3 and 4 consider the main issues concerning carbon trading and taxation. Section 5 compares both instruments and offers a synthesis of the main policy arguments. Finally, sections 6 and 7 offer some policy guidance and some practical conclusions regarding the issue at stake: the optimal instruments for effective climate change policy.

2. - Market Based Instruments

Welfare economics shows us that self-interested individuals in a well-functioning market economy are guided by "an invisible hand" to an efficient outcome, that is, an outcome where it is not possible to increase the welfare of one or more individuals without harming someone else's welfare (Smith, 1776). However, challenges such as global warming incorporate "market failures" that prevent the achievement of efficient outcomes and thus force governments to regulate externalities and generate incentives to correct the under-provision of greenhouse gas abatement effort (Sandler, 1997).

One way to deal with inefficient levels of greenhouse gas (GHG) emissions or any other negative externality is to use a "command and control" fix, by prohibiting or at least limiting emissions. However, direct regulation of polluting activities only makes sense in some very specific cases. In many situations command and control measures result in significant opportunity costs due to the lack of flexibility or lack of capacity to exploit opportunities derived from this type of interventions.

A preferred alternative for most economists to deal with externalities is to rely on a market-based approach, where agents are given, *via* prices, an incentive to reduce GHG emissions. There are at least two classes of market-based instruments here: taxes and permit trading systems. An emission tax is a price mechanism in its purest sense, as a "Pigovian" price is fixed and the emission level is allowed to vary according to economic activity. The permit trading systems is, in contrast, a quantity mechanism, as it fixes the overall quantity of emissions and allows the price to vary. In an ideal setting, both types of instruments are more or less equivalent: the price of an emission-permit produces the same incentives to reduce emissions as a Pigovian tax (Baumol and Oates, 1988). However, there are political economy issues as well as reasons related to the different types of uncertainty present in global warming that may explain why policy-makers are not indifferent with regard to price *versus* quantity market mechanisms (Weitzman, 1974; Markandya, this issue).

In the real world, Finland pioneered CO_2 pricing by the implementation of the first global CO_2 tax in 1992 by taxing energy products (Labandeira *et* al., 2008). Experiences such as the CO_2 tax in Norway or Denmark (Labandeira *et* al., op. cit.; World Bank, 2010) as well as the Climate Change Levy in the UK (OECD, 2004) are further good examples. The impact of the taxing schemes has not always been as positive as expected by some authors, but in general terms it can be recognised to be a very useful policy instrument (Newell and Pizer, 2003). With regard to marketable emission permit systems, their use is limited, but has undergone considerable growth in recent years. Experience in the US with markets for some air pollutants has been evaluated very positively (Ellerman *et* al., 2000) and in June 2005, greenhouse gas emissions trading emerged in the European Union through the so-called European Union Emissions Trading Scheme (EU ETS, here after). The following two sections consider the performance of these two types of market-based instruments.

3. - Carbon Trading

The theory of carbon trading is fairly simple. A maximum limit ("cap") is established on the carbon emissions for participants in the scheme and emission permits are issued for each tonne of emissions allowed under the cap. The permits are then distributed to members of the scheme either through auctioning or "grandfathering" (free distribution). If the cap is constraining, then polluters will be forced to either reduce their emissions or buy reductions in emissions from other participants. Trade is introduced, as explained in the previous section, to exploit opportunities, by allowing the members who find cutting carbon emissions to be expensive to buy permits from those who have cheaper abatement options[1]. This ensures that the emission reduction is achieved at the lowest cost possible.

The EU ETS is the best example so far for an emission trading market and a proto-type for an eventual global GHG emissions trading system (Aldy and Stavins, 2008). The EU ETS Directive (EC, 2003) mandated an initial three-year trading period for 2005-2007, often called the pilot or trial phase, to be followed by a second five-year trading period for 2008-2012 that corresponds to the First Commitment Period under the Kyoto Protocol, and subsequent *post*-2012 trading periods. There is extensive literature that analyses the performance of the EU ETS (Ellerman and Buchner, 2007; Ellerman and Joskow, 2008 and Convery, Ellerman and De Perthuis, 2008). We will here use the EU ETS experience to discuss relevant aspects of the design and performance of a carbon trading system.

The performance of a cap-and-trade system has at least two critical dimensions that we should look at: *(1)* its ability to provide a single "Pigovian" price and *(2)* the trading volume generated. Additionally, the emission permit allocation process is also an important feature to determine the distributional fairness and even the dynamic efficiency of the cap-and-trade system.

---

[1] The market may also be enriched by other flexibility instruments such as offset credits from abroad and joint implementation projects.

3.1 *Price*

If a cap-and-trade system is going to lead to the least-cost attainment of the emission constraint, then every agent taking part in the market need to face the same signal and this signal has to reflect the marginal cost that can be economically justified when reducing emissions and adjusting production. Graph 1 displays the evolution of emission allowance prices in the EU ETS from 2005 to 2010[2].

The evolution of the emission allowance prices in the EU ETS presents the following features. First, the price during the so-called "trial period" (2005-2007) was higher than initially expected. Second, the 2006 price adjustment was very abrupt, especially for trial period allowances. Third, there was some volatility in evolution of the price. All these three features, however, will be easily explained below and do not mean that the EU ETS fails to generate a reliable market price for CO_2 emissions.

Graph 1

EVOLUTION OF ALLOWANCE PRICES IN THE EU ETS

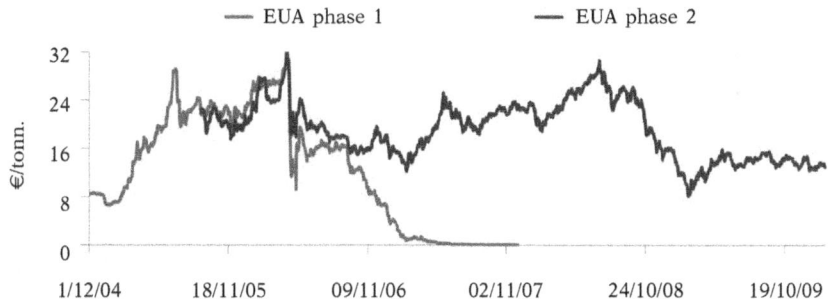

Source: Point Carbon (2010).

---

[2] Note that the absence of banking from the so-called trial period (2005-2007) to the second period (2008-2012) makes the futures contract to deliver in December 2007 and the futures contracts to deliver in December 2008, 2009 and 2010 different products with different prices.

As Ellerman and Joskow explain, initial expectations about prices are often wrong in many cap-and-trade systems. This is partly due to uncertainty regarding the effect on the demand for allowances of unpredictable variables such as economic activity, weather and energy prices, and also due to the lack of knowledge on how abatement activities will respond to the new emission price. Ellerman *et al.* show that the initial emission reports in the US SO_2 trading systems also revealed lower emissions than expected and that the resulting adjustment in expectations made the price of allowances decline significantly. However, the expectations were adjusted more rapidly in the case of the EU ETS than for the US SO_2 trading system. The explanation for this sharp adjustment in the case of the first period emission allowances in the EU ETS (EUA Phase 1 in Graph 1) is straightforward. It was due to the fact that the trial trading period was designed as a self-contained process without the possibility of 'banking' emissions for the second phase[3]. Therefore, the adjustment had to be performed over a truncated time-horizon, ending in December 2007. In the case of US SO_2 emission allowances and even of EU ETS second-phase emission allowances (EUA Phase 2 in Graph 1), this adjustment could be spread over a longer period of time and led to a fairly smooth transition. With regard to price volatility, Ellerman and Joskow argue that these price movements are not unusual for cap-and-trade systems and they are comparable to the movements of related energy commodities. Obviously, there are improvements such as the introduction of banking and increasing the emission data reporting frequency, that would reduce price volatility and, therefore, enhance the effectiveness of allowance prices for providing reliable incentives for abatement. Nevertheless, the ability of the EU ETS to generate a visible and unique market price for emissions should not be called into question.

---

[3] Banking refers to the possibility of carrying over an assigned amount of emission allowances from one compliance period to the following period in anticipation that these allowances will accrue value over time or to hedge against future price changes.

3.2 *Quantity*

Prices are one side of the coin of market performance. The other side of the coin is the trading volume. Graph 2 shows how the trading volume in the EU ETS has steadily grown since its creation in 2005[4]. The quantity of allowances exchanged in 2005 was relatively low at 262 Mt, but the quantity increased nearly fourfold in 2006 and has continued growing ever since.

GRAPH 2

EVOLUTION OF TRADING VOLUME IN THE EU ETS

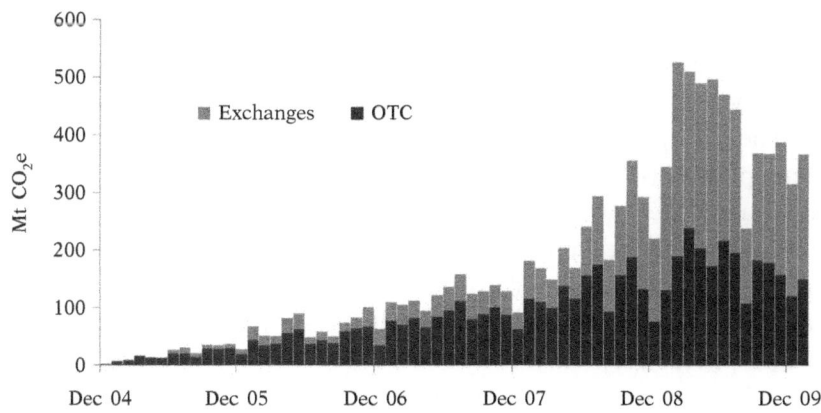

Source: POINT CARBON (2010).

One of the risks when designing a market is over-allocation of permits (resulting in a non-constraining cap), which will prevent substantial trading from taking place. The expected emission reduction for the EU25 as a whole during the trial period was not higher than two percentage points. Thus, it is not surprising that some member states and sectors received allocations that were larger than their expected and actual emissions. This might have had the undesired effect of insufficient emission trade and abatement effort during the trial period. Ellerman and Buchner esti-

---

[4] Bilateral forward trades for allowances began in the spring of 2003, well before the official start of the EU ETS in January 2005.

mate that abatement in 2005 and 2006 was probably between 50 and 100 millions tons/year (2%-5% of covered emissions). However, Ellerman and Joskow stress that the primary goal of the EU ETS during the trial period was not to effect significant emission reductions but to develop the market infrastructure and that the more ambitious emission reduction targets in the second phase will probably result in more trade and abatement effort.

3.3 *Allocation Process*

The distribution of allowances by member states to cover installations is the third key element that might merely be seen as a mechanism to deal with distributional justice, but which could certainly distort abatement behaviour or competition.

The EU ETS allocation process has evolved through its different phases (Grubb *et* al., 2009). Thus, Phase 1 allowances were distributed free-of-charge and some sectors, particularly power companies, enjoyed enormous windfall profits. In Phase 2, free allowances to power generators were cut more than other sectors. Phase 3 seems to be leading to full auctioning in the power sector and protective measures to sectors at risk of "carbon leakage"[5]. There is a growing interest in determining how big the carbon leakage problem can be and how can preventive policies be implemented. Some of the preventive policies suggested include cost containment measures, international sectoral agreements, allocation rebates or border adjustment (Fischer and Fox, 2009; Asselt and Biermann, 2007; Grubb and Neuhoff, 2006; Biermann and Brohm, 2005; Kuik and Gerlagh, 2003; Neuhoff, 2008).

Economic theory shows that the allocation method should not affect the economic efficiency of an emission trading market (Montgomery, 1972). In certain ideal circumstances (small programs for local pollutants), incentives for economic agents to cut

---

[5] Carbon leakage occurs when, due to asymmetries in climate policy, actions designed to cut GHG emissions implemented in one country/jurisdiction, lead to the shifting of the targeted emitting activities elsewhere, thus undermining the attempt to reduce emissions.

emissions remain the same even if they receive free allowances. Thus, there is a tendency to choose free allocation as a way of addressing competitiveness concerns from industry.

Yet we should not forget that the pricing of CO_2 emissions does not fit in the "ideal circumstances" described above. The market for emission allowances is not a small programme, and therefore its potential revenues should be seriously considered as opportunities for displacing other burdensome taxes. Furthermore, CO_2 is a global pollutant, and in the absence of a global carbon market, carbon leakage to unregulated sectors and countries can limit the effectiveness of the cap-and-trade system. Table 1 offers a summary of the perverse incentives arising from free allowance allocation, which implies a compelling economic rationale to choose auctioning as an allocation method.

TABLE 1

PERVERSE INCENTIVES ARISING FROM FREE ALLOWANCE ALLOCATION

Method of Allocation	Distortions	Plant Operation		Investment in plants		Energy efficiency	
		Dirtier plants	Encourages operation	Discourage plant closure	Dirtier plants	Reduces incentives for consumers	Reduces incentives for producers
Benchmarking	Capacity			✓			
	Capacity by fuel/plant type			✓	✓		
Updating from previous periods	Output		✓	✓		✓	
	Output by fuel/plant type	✓	✓	✓	✓	✓	
	Emissions	✓	✓	✓	✓	✓	✓
Output-based undifferentiated allocation	Final product		✓	✓		✓	
	Intermediate product		✓	✓		✓	✓

Note: Benchmarking is an allocation system based on product-specific benchmarks. In this case, an installation receives a predetermined amount of allowances for each unit of the good that it produces. The level of this benchmark should be set so that only the best-in-class receive a full supply of allowances. All other installations then must either improve their efficiency to the benchmark level, or they need to purchase additional allowances if efficiency improvements are not feasible. Updating from previous periods involves updating allowance allocations based on emission performance in the previous allocation periods. In an output-based allocation system installations would receive permits proportional to their share of their industry's output.

Source: adapted from NEUHOFF K. (2008) and GRUBB M. *et al.* (2009).

4. - Carbon Tax

As it has already been mentioned, a properly designed carbon tax system can also be a cost-effective policy instrument for environmental protection (Baranzini *et* al., 2000; World Bank, 2005). In any event, the debate on the "double dividend" (Pearce, 1991)[6] and the lack of strong empirical evidence to support it (Bovenberg and de Mooij, 1994) indicates that environmental taxation faces many difficulties in the design and implementation phase that should be properly addressed. A profound analysis on the impacts of Green or Environmental Fiscal reforms in Europe can be found in Skou and Ekins (2009) where they looked at the use of carbon and energy taxes to reduce other taxes that might distort the economy.

A carbon tax is an excise duty defined by the carbon content of fossil fuels. In some cases, the carbon tax might be approximated through energy taxes (a fixed amount per unit of energy *i.e.* kilowatt-hour) as the latter indirectly also fixes a price over CO_2 emissions. However, it should be noted that carbon taxes and energy taxes are essentially different. For instance, a carbon tax will not be incurred by nuclear energy while an energy tax will do (Zhang and Baranzani, 2009).

But let us describe *the main issues arising* when discussing the use of carbon taxes:

The *first issue* to be highlighted is that there is only limited experience in the use of carbon taxes and thus empirical evidence regarding the impacts is limited. Carbon taxes in some form have been implemented in developed countries such as Norway, Sweden, the Netherlands, Denmark, Finland, Italy, Switzerland or the UK. Good descriptions of the tax systems in each country can be found in Skou and Ekins; Labandeira *et* al. (op. cit.); Pearce (2005) and European Commission (1999). The UK has recently set a special independent commission to look at the Green Fiscal Reform

---

[6] The double dividend hypothesis refers to the idea that a revenue-neutral substitution of an environmental tax for other existing income taxes can contribute to protect the environment as well as reducing other economic costs derived from distortions existing due to taxation.

in depth. The "Green Fiscal Commission" published its final report in 2009 concluding that carbon tax is extremely adequate instrument for climate policy with strong public support for its use[7].

Secondly, taxes not originally intended to affect emissions also create charges on CO_2. In other words, energy tax and other duties on energy sales generated are responsible for what is known as the "implicit carbon tax" which differs significantly over countries and therefore the average price of a ton of CO_2 is also very different (Baranzini *et* al.). If we add to this the effect of subsidies over other energy sources or energy uses, it is easy to understand why the use of a global carbon taxes is often questioned (Baranzini *et* al.).

Thirdly, the fact that if no redistribution of revenues is foreseen, carbon taxes can be regressive as well as impose higher costs to polluters than other command and control policies. This usually leads to the discussion of whether the revenues should be *earmarked* to finance specific environmental protection programs, *devoted to compensate* the most affected population for the regressivity or contribute to *revenue neutrality* (Green Fiscal Reform) by reducing other distorting taxes (Bovenberg and Goulder, 2002; OECD, 1996; Galarraga and Gonzalez-Eguino, 2005).

Fourthly, there is extensive literature devoted to analysing the impact of carbon taxes on Competitiveness, the Distributive impact and the Environmental impact. Baranzini *et* al. conclude that potential impact on competitiveness is not significant according to existing literature. Likewise the tax is not as regressive as expected and studies usually do not account for environmental benefits attached to the use of taxes (Gough *et* al., 2008). In any case, the analysis from these three perspectives is greatly conditioned by the discussion on the use of the collected revenue in the preceding paragraph. Other studies such as Skou and Ekins also support these conclusions.

---

[7] *http://www.greenfiscalcommission.org.uk/index.php/site/about/final_report/*

5. - Tax *vs.* Trading

In the recent years, the literature has focussed on comparing the use of the two policy instruments presented in this paper: carbon trading and carbon tax (Skou and Ekins). The starting point of this discussion is that under certain conditions both instruments can be designed and implemented in such a way that the outcome in terms of efficiency could very similar. Parry and Pizer (2007) offer a very interesting insight and argue that the main differences can be summarised as follows:

a. Certainty over price: carbon taxes are market-based instruments designed to set the price at which the ton of carbon could be emitted and, as such, offer a great degree of certainty regarding the price of carbon. Carbon trading instead focuses on limiting the quantity of CO_2 tons emitted and thus pays little attention to controlling the prices. During the first period, the EU-ETS experienced significant price volatility attributed to the grandfathering principle, the generous distribution of allowances and other problems arising from incomplete data and other intrinsic emission projection issues (Grubb *et* al.). Designing features such as safety valves or borrowing could significantly reduce price volatility (Parry and Pizer).

b. Certainty over emissions: in terms of this characteristic, carbon trading achieves better results as it sets the rights according to the emission targets planned for that period. In fact, the EU ETS has cut emissions in the EU by 120-300 million metric tonnes of carbon dioxide during the first phase (5% of emissions by the sectors affected by Directive 2003/87/EC). The carbon tax instead will not be able to set this limit as the quantity of tons emitted will depend upon energy demand and other factors that might affect the energy price.

c. Efficient encouragement of least cost emissions reductions: Available literature suggests that both systems are capable of encouraging least cost emissions. In fact, "in a world where emissions externality is the only market distortion and there is no uncertainty, either instruments could achieve the first best outcome, if the emission cap at each date equals the emissions that would result under the Pigovian tax" (Aldy *et* al., 2009). The trading and tax system differ substantially when other tax distortions exist, uncertainty is present and distributional impact is important. A well-structured defence of the carbon tax system over the trading schemes in terms of efficiency and effectiveness can be found in Green *et* al. (2007).

d. Ability to raise revenue: This property is traditionally only attributed to tax systems but the truth is that this can be offset by incorporating auctioning schemes in the cap and trade systems. Directive 2009/28/EC currently incorporates auctioning in the EU ETS progressively, 20% by 2013, 70% by 2020 and 100% by 2027 contributing to reducing other distortions such as windfall profits (Gallastegui and Galarraga, 2010).

e. Incentives for R&D: The carbon tax is in principle the one that enhances dynamic incentives as it is highly effective to generate incentives over time to invest in R&D to reduce emissions. This is in fact one of the most appreciated attributes of environmental taxes. Although trading schemes could be designed to generate some significant incentives for innovation, the fact is that expectations regarding the CO_2 permit prices will directly affect these incentives.

f. Harm to competitiveness: both systems can harm international competitiveness and this has been extensively analysed in the literature (Hauser *et* al., 2008; Hourcade *et* al., 2008; Smale *et* al., 2006). Yet the evidence is that in the case of EU ETS, the impacts have been limited to a number of industry sectors that could be accounted for through complementary measures. This might be a serious problem for Old Industrial Regions where there is a high concentration of vulnerable industrial sectors (Gonzalez-Eguino *et* al., 2010). In fact, Directive 2009/29/CE requires changes in the EU ETS to prevent the so-called carbon leakage and pays special attention to vulnerable sectors. The use of the raised revenue and any income generated through auctioning in trading schemes seem to be necessary to offset these impacts. Other tailored solutions will be necessary for exposed sectors (Grubb *et* al.).

g. Practical barriers to implementation: It is well known that new taxes always face very strong opposition and environmental taxes are no exemption. Both the US and EU have been postponing the debate on new fiscal instruments to support climate policy for many years now (or even decades). Policymakers seem to be more receptive towards permit trading schemes. However, carbon trading faces opposition from affected industry and other economic sectors and it is particularly sensitive to allocations *via* auctions.

h. New institutional requirements: In fact, the trading scheme requires stronger institutions to be set up to be able to trade the rights on the market, to verify emissions and other related supportive actions. But the truth is that these institutions have arisen very quickly, effectively and at very reasonable costs in trading scheme experiments.

Other issues to take into account when comparing both systems, according to Green *et* al., are: the fact that carbon taxes can support the elimination of other superfluous regulations such as some fuel standards; the possibility to adjust the tax to be more or less stringent, thus allowing the markets to react with certainty; and allowing the revenues collected by the tax to stay in the country and thus reduce the opposition in the international arena. The latter could also enable specific programs to be designed to support those most vulnerable sectors.

6. - Lessons to be Learnt

There is no clear consensus regarding the superiority of one instrument over the other as the literature shows that both instruments could be designed in such a way, at least theoretically, that lead to similar outcomes. Goulder (2009) deals with this issue and insists that the choice *per se* is not the issue but the fact that the instrument chosen is well designed. Of course, both can represent different trade-offs between efficiency, distributional impacts and difficulties for implementation.

Conservative and more liberal parties in many countries are embroiled in an intellectual (political) discussion over whether the Government should intervene in the economy or not. A good example of these discussions is the case of the Waxman-Markey Act that was recently discussed at the US Senate, where contradictions seem to be flourishing among the defenders and those opposing the instrument (Krugman, 2010).

In any case, it could be stated that there is some consensus among policy makers and economists on the usefulness of pricing (GHG) emissions as an effective climate policy instrument. Neuhoff clearly states that «countries where energy prices are high require less energy per unit of GDP than countries with low energy prices. This suggests that carbon pricing is an essential policy instruments to deliver emissions reductions».

Of course, different schools of economists both in the EU and US are leading the debate on the superiority of one instrument over the other. Green *et* al., for instance, argue that «a cap and trade approach to controlling GHG emissions [globally] would be highly problematic» due to a lack of an international binding authority and the great incentives for cheating. Nordhaus (2006) strongly supports this idea by adding that a price approach gives less room for corruption as it does not create artificial scarcities and monopolies.

Other authors such as Aldy *et* al. also reveal some potential benefits of enhanced cap and trade systems while Grubb *et* al. clearly put forward as their first idea that *emission trading works*. Parry and Pizer suggest that an enhanced hybrid trading system could be very effective. This is a scheme in which a "safety valve" exists in the form of an amount of rights that can be sold by the government to increase supply if prices are too high and allow banking to prevent prices from going too low. That is, if rights can be kept for future phases of the trading scheme, the expected value of them will increase as explained in section 3. In other words, when due to a generous initial allocation of permits (as it happened in the first phase of the EU ETS 2005-2007) or due to any other reason the demand for rights is limited for that period, if banking is allowed the expectations of future shortage of permits will prevent prices from collapsing. In a far more technical paper, Newell and Pizer conclude that «price-based instruments for carbon reduction – such as a carbon tax – are likely to generate several times the expected welfare gains of quantity-based instruments, such as tradable carbon permits. Yet, programs such as the Kyoto Protocol require binding, quantity-based reductions. This suggests that there could be at least great value in incorporating price elements into quantity-based policies, such as a cap on emission permit prices». This conclusion is based on the analysis developed in Weitzman regarding the effectiveness of instruments in relation of the slopes of marginal benefits curves. In this case authors incorporate a comparison with a measure of marginal benefit that accounts for growth, discounting, depreciation and correlation of cost shocks.

In terms of political feasibility, there is a clear bias towards carbon markets over taxation schemes. In fact, while the carbon tax has been long and deeply discussed in EU institutions, carbon trading was discussed, approved and set up reasonably fast in very short time. This includes the new Directive 2009/28/EC on carbon trading as well as the institutions created with the flexibility mechanisms of the Kyoto Protocol.

Neuhoff offer a clear explanation of the role of expectations when choosing between one or other instrument, stating that «short-term emissions reduction targets are typically based on assessments of available technologies and might lend themselves to carbon taxes» while «long-term targets reflect the need to avoid extreme climate impacts and the expectations about technology improvements and innovation» and thus will justify the use of a carbon trading scheme. In any case, emission trading could be an option to reconcile both visions by offering a «reserve price in auctions or government issued put options» to protect from overly low carbon prices while assuring a credible emission reduction target to reduce uncertainty.

Some important recommendations that could be highlighted from the literature are:

– Carbon pricing should look at the advantages of trading schemes aimed at emission reductions at lower costs, but should also take into consideration other factors such as cultural preferences, transactions costs and other management implications. (Neuhoff).

– Although awareness raising should also be an important part of climate policy, we cannot rely solely on altruism and a system that generates private incentives for fewer emissions should be sought (Krugman).

– Carbon pricing can generate revenues that could be used to promote low carbon technologies or to reduce other distorting taxes. This might be of great help to reduce the distributional impact of the system and to seek greater political feasibility, even though taxation is a word policy makers do not like using!

– Committing to emission targets at different levels of ambition to reflect the domestic awareness and support might be seen as a very practical and realistic approach to overcome the constraints on approving a second part for the Kyoto Protocol. This is better than waiting for an ambitious harmonised approach. (Neuhoff).

– Internationally differing carbon prices could be handled and carbon leakage only represents a major issue for some few sectors. This might need tailored solutions in the form of special aid or sectoral agreements (Neuhoff; Grubb, *et* al.; Gonzalez *et* al.).

– An emission trading scheme could be developed that minimises the negative effects of EU ETS, which is a very effective policy instrument as plenty of expertise has been developed in the area (Grubb, *et* al.).

– Price volatility might be an issue to be accounted for and is a weakness in trading schemes. Thus instruments should be developed to minimise it. Improving emissions projections could help.

Systems such as border adjustments should only be used for situations in which carbon leakage is a problem and under strict multilateral agreement so as not to affect international trading (Grubb *et* al.). In fact, as well as short term leakage, long term leakage should also be taken into account (Goldemberg *et* al.). This refers to the fact that most of the studies look at situations in which carbon-intensive production moves to areas in which regulations is less stringent through studies based "on standard international trade models, in which the location of production of various goods and machines is fixed" and therefore estimations are based only on commodity substitution. However, in the long run carbon leakage effects might be greater as relocation of industries might happen (Goldemberg *et* al.).

The debate could be summarised as follows: while there exists a clear consensus towards the need for carbon pricing, there seems to be a divergence in the literature with regard to the effectiveness of the proposed instruments (trading *versus* tax). The possibility of moving towards hybrid trading schemes, where "safety valves" are allowed and auctioning limited, offers many advantages over any carbon tax system. Other factors, such as political feasibility, idiosyncrasy and awareness of the population in each country, should also be taken into account when choosing one instrument. The chosen scheme being well designed is more important than the nature of the instrument itself.

7. - Conclusions

The discussion presented in this paper is a starting point to consider many issues regarding CO_2 pricing as an effective mechanism of climate policy. The two main instruments used so far have been described herein: carbon tax and carbon trading. Both can provide equivalent results if applied under certain conditions. The main feature has to do with the fact that, irrespective of the instrument chosen, it has to be properly designed. In particular current distorting taxes such as energy taxation or other fiscal instruments should be carefully taken into account and acknowledged to minimise undesired effects.

Other issues such as the political feasibility of implementing one instrument over another should be always taken into account. Policy makers often prefer not to have proposed new taxes unless they can be very well designed and justified. The use of the tax revenue collected provides a great opportunity to reduce negative impacts and opposition.

Many authors provide stimulating discussions on the benefits of one instrument over the other and many insights to help to fine tune the chosen policy instrument. The most recent literature seems to devote some time to the idea of hybrid systems. The conclusion by Newell and Pizer presented in the previous section seems to strongly support this recommendation.

In fact, when one looks at the distribution of emissions among activity sectors, the main sources of emissions are often energy production, transport and industry. One could argue that industry and energy producing sector in the EU is already doing a big effort in terms of internalising the carbon price though the ETS. However, other sources of energy consumption still need to be encouraged to share the burden of carbon reduction. This might lead into the discussion on how mitigation efforts should be distributed among sectors but also among countries and regions, *i.e.* setting targets in terms of emissions *per capita* or emissions per unit of output. That is, shall we consider emissions for unit of output produced or shall we defend the view that countries should face the burden in terms of the number of inhabitants? This is in fact a very sensitive and political issue and at the core of the UNFCCC meetings discussions.

One possible half way approach is proposed in Galarraga (2010), that consists in designing a system that accounts for *per capita* emissions for sectors responsible for diffuse emissions (such as the residential sector or transport) while continuing with a emissions per unit of GDP for other industrial and economic activities. This could lead the policy makers towards a hybrid system in which carbon trading is used for well regulated and controlled economic activities (similar to the enhanced EU ETS) while a carbon tax fixed in terms of *per capita* emissions is chosen for other diffuse emissions. It could allow the focus to be simultaneously on equity and efficiency issues (Galarraga)[8].

The choice of the instrument depends on the particular context of each governing body in each country and the development of the ongoing discussions for global CO_2 emission cuts that will be highlighted in the next Cop 16 in Cancun (Mexico). Using ei-

---

[8] Equity refers to the sharing of the burden among all people and countries in similar basis whereas efficiency refers to the fact that emission reductions take place wherever it is cheaper to abate.

ther a single global or many national carbon taxes to reinforce the actual flexible mechanisms of the Kyoto Protocol (*i.e.* Carbon Market, Clean Development Mechanisms and Joint Implementation) that have already delivered some results could be an excellent solution from the point of view of economic efficiency, distributional fairness and political feasibility.

BIBLIOGRAPHY

ALDY J.E. - KRUPNICK A.J. - NEWELL R.G. - PARRY I.W.H. - PIZER W.A., «Designing Climate Mitigation Policy», *RFF Discussion Paper*, 08-16-REV, 2009.

ALDY J.E. - STAVINS R.N., «Introduction: International Policy Architecture for Global Climate Change», in ALDY J.E. - STAVINS R.N. (eds.), *Architectures for Agreement: Addressing Global Climate Change in the Post-Kyoto World*, Cambridge and New York, Cambridge University Press, 2008.

BARANZINI A. - GOLDEMBERG J. - SPECK S., «A Future for Carbon Taxes: Survey», *Ecological Economics*, Vol. 32, 2000, pages 395-412.

BAUMOL W.J. - OATES W.E., *The Theory of Environmental Policy*, 2nd edition, Cambridge University Press, 1988.

BIERMANN F. - BROHM R., «Implementing the Kyoto Protocol Without the United States: The Strategic Role of Energy Tax Adjustments at the Border», *Climate Policy*, vol. 4, 2005, pages 289-302.

BOVENBERG A.L. - DE MOOIJ R., «Environmental Levies and Distortionary Taxation», *American Economic Review*, Vol. 84, 1994, pages 1085-1089.

BOVENBERG A.L. - GOULDER L.H., «Environmental Taxation and Regulation», in AUERBACH A. - FELDSTEIN M. (eds.), *Handbook of Public Economics*, Vol. 3, Elsevier Science, North-Holland, 2005.

CONVERY F. - ELLERMAN A.D. - DE PERTHUIS C., «The European Carbon Market in Action: Lessons from the First Trading Period: Interim Report» (March), 2008, available at *http://www.aprec.fr/documents/08-03-25_interim_report_en.pdf*.

ELLERMAN A.D. - BUCHNER B.K., «The European Union Emissions Trading Scheme: Origins, Allocation, and Early Results», *Review of Environmental Economics and Policy*, Vol. 1, 2007, pages 66-87.

ELLERMAN A.D. - JOSKOW P.L., *The European Union's Emissions Trading System in Perspective*, Report prepared for the Pew Center on Global Climate Change, 2008.

ELLERMAN A.D. - JOSKOW P.L. - SCHMALENSEE R. - MONTERO J.P. - BAILEY E., *Markets for Clean Air: The US Acid Rain Program*, Cambridge University Press, 2008.

EUROPEAN COMMISSION, *Database on Environmental Taxes in the EU Member States plus Norway and Switzerland - Evaluation of Environmental Effects of Environmental Taxes*, Office for Official Publication of the European Communities, Luxembourg, 1999.

— - —, *Directive 2003/87/EC of the European Parliament and of the Council of 13th October 2003 establishing a scheme for greenhouse gas emissions allowance trading within the Community and amending Council Directive 96/61/EC*, 2003.

FISCHER C. - FOX A.K., «Comparing Policies to Combat Emissions Leakage: Border Tax Adjustments versus Rebates», *RFF Discussion Paper*, no. 09-02, Washington, 2009.

GALARRAGA I., «A Discussion of "Involving Developing Countries in Global Climate Policies" by Anil Markandya», in CERDA E. - LABANDEIRA X. (eds.), *Climate Change Policies: Global Challenges and Future Prospects*, Edward Elgar Publishing, 2010.

GALARRAGA I. - GONZÁLEZ-EGUINO M., «La fiscalidad ambiental: un nuevo reto, una nueva oportunidad», *Zergak, Gaceta Tributaria del País Vasco*, ISSN 1133-5130, 2005, pages 133-147.

GALLASTEGUI M.C. - GALARRAGA I., «La Unión Europea frente al cambio climático: el paquete de medidas sobre cambio climático y energía (20-20-20)», in BECK-

ER F. - CAZORLA L.M. - MARTÍNEZ-SIMANCAS J. (eds.), *Tratado de energías renovables*, Iberdrola-Thomson-Aranzadi, 2010.

GREEN K.P. - HAYWARD S.F. - HASSETT K.A., «Climate Change: Cap vs. Taxes», *Environmental Policy Outlook*, American Enterprise Institute for Public Policy Research, AEI, no. 2, June, 2007.

GOLDENBERG J. - SQUITIERI R. - STIGLITZ J. - AMANO A. - SHAOXIONG X. - SAHA R., «Introduction: Scope of the Assessment», in BRUCE J.P. - LEE H. - HAITES E.F. (eds.), *Climate Change 1995: Economic and Social Dimensions of Climate Change*, Contribution of Working Group III to the Second Assessment Report of the Intergovernmental Panel on Climate Change, Cambridge University Press, Cambridge, UK and New York, US, ISBN 9780521568548, 1996.

GONZÁLEZ-EGUINO M. - GALARRAGA I. - ANSUATEGI A., «Carbon Leakage and the Future of Old Industrial Regions after Copenhagen», *BC3, Working Paper*, Series 2010-02, Basque Centre for Climate Change (BC3), Bilbao, Spain, 2010.

GOUGH I. - MEADOWCROFT J. - DRYZEK J. - GERHARDS J. - LENGFELD H. - MARKANDYA A. - ORTIZ R.A., «JESP Symposium: Climate Change and Social Policy», *Journal of European Social Policy*, Vol. 18, 2008, pages 353-365.

GOULDER L.H., «Carbon Taxes versus Cap and Trade», *Working Paper*, Department of Economics, Stanford University, 2009.

GRUBB M. - BREWER T.L. - SATO M. - HEILMAYR R. - FAZEKAS D., «Climate Policy and Industrial Competitiveness: Ten Insights from Europe on the EU Trading System», *Climate and Energy Paper Series*, GMF, 2009.

GRUBB M. - NEUHOFF K., «Allocation and Competitiveness in the EU Emissions Trading Scheme: Policy Overview», *Climate Policy*, Vol. 6, no. 1, 2006, pages 137-160.

HAUSER T. - BRADLEY R. - CHILDS B. - WERKMAN J. - HEILMAYR R., *Leveling the Carbon Playing Field: International Competition and US Climate Policy Design*, Peterson Institute for International Economics, World Resource Institute, Washington, DC, 2008.

HOURCADE J.C - NEUHOFF K. - DEMAILLY D. - SATO M., «Differentiation and Dynamics of EU ETS Industrial Competitiveness Impacts», *Climate Strategies Papers*, 2008.

KUIK O. - GERLAGH R., «Trade Liberalization and Carbon Leakage», *The Energy Journal*, Vol. 24, 2003, pages 97-120.

KRUGMAN P., «Building a Green Economy», *The New York Times*, 5th of April, 2010, available at *http://www.nytimes.com/2010/04/11/magazine/11Economy-t.html*.

LABANDEIRA X. - LÓPEZ OTERO X. - RODRÍGUEZ MÉNDEZ M., «Cambio climático y reformas fiscales verdes», *Ekonomiaz*, Vol. 67, 2008.

MONTGOMERY W.D., «Markets in Licences and Efficient Pollution Control Programs», *Journal of Economic Theory*, Vol. 5, no. 3, 1972, pages 395-418.

NEUHOFF K., *Tackling Carbon: How to Price Carbon for Climate Policy*, Report by Climate Strategies, Faculty of Economics, University of Cambridge, 2008.

NEWELL R.G. - PIZER W.A., «Regulating Stock Externalities Under Uncertainty», *Journal of Environmental Economics and Management*, Vol. 45, 2003, pages 416-432.

NORDHAUS W.D., «After Kyoto: Alternative Mechanisms to Control Global Warming», *AEA Papers and Procedings*, Vol. 96, no. 2, 2006.

OECD, *Implementation Strategies for Environmental Taxes*, OCDE, Paris, 1996.

— - —, *The United Kingdom Climate Change Levy: A Study in Political Economy*, COM/ENV/EPOC/CTPA/ CFA, 2004 66/FINAL.

PARRY I. - PIZER B., «Combating Global Warming: Is Taxation or Cap-and-Trade The Better Strategy for Reducing Greenhouse Emissions», *Regulation*, Vol. 30, no. 3, 2007.

PEARCE D., «The Role of Carbon Taxes in Adjusting Global Warming», *Economic Journal*, Vol. 101, 1991, pages 38-948.

— - —, «The Political Economy of an Energy Tax: The United Kingdom's Climate Change Levy», *Energy Economics*, Vol. 28, 2005, pages 149-158.

POINT CARBON, «Carbon 2010: Return of the Sovereign», in TVINNEREIM E. - ROINE S. (eds.), *Point Carbon*, 2010.

SANDLER T., *Global Challenges: An Approach to Environmental, Political, and Economic Problems*, Cambridge University Press, 1997.

SKOU A. - EKINS P., *Carbon-Energy Taxation: Lessons from Europe*, Oxford University Press, 978-0-19-957068-3, 2009.

SMALE R. - HARTLEY M. - HEPBURN C. - WARD J. - GRUBB M., «The Impact of CO_2 Emissions Trading on Firm Profits and Market Prices», *Climate Policy*, Vol. 6, no. 1, 2006, pages 31-48.

SMITH A., *The Wealth of Nations: An Inquiry into the Nature and Causes*, printed for Strahan W. – Cadell T., London, 1776.

STERN N., *Stern Review on the Economics of Climate Change*, Executive Summary, HM-Treasury, 2006.

VAN ASSELT H. - BIERMANN F., «European Emissions Trading and the International Competitiveness of Energy-Intensive Industries: A Legal and Political Evaluation of Possible Supporting Measures», *Energy Policy*, Vol. 35, no. 1, 2007, pages 497-507.

WEITZMAN M.L., «Prices versus Quantities», *Review of Economic Studies*, Vol. 41, no. 4, 1974, pages 477-491.

WORLD BANK, *World Bank Development Report 2010: Development and Climate Change*, The World Bank, Washington DC, 2010.

— - —, *Environmental Fiscal Reform: What Should be Done and How to Achieve It*, The World Bank, Washington DC, 2005.

ZHANG Z.X. - BARANZINI A., «What Do We Know About Carbon Taxes? An Inquiry into Their Impacts on Competitiveness and Distribution of Income», *MPRA Paper*, no. 13225, 2009.

Green Taxes on Aviation: The Case of Italy. The Proposal of the Green Taxation Matrix

Anil Markandya   -   **Elena Claire Ricci***

University of Bath
Ikerbasque Professor, Basque Centre
for Climate Change

University of Study Milan
Fondazione Eni Enrico Mattei

This work aims at analysing the impacts of an aviation tax in Italy and at proposing a general methodology to manage the complex effects of introducing environmental taxes. After a review of the literature and the analysis of the possible structures of the tax, its effects, at various levels, are calculated under a partial-equilibrium model. These are then introduced in an Input-Output model to evaluate the impacts on the other sectors of the economy. Finally, different revenue-recycling options are discussed. We propose as an instrument of comparison the Green Taxation Matrix (GTM) that can act as a guide for policy-makers. [JEL Classification: D57, H23, Q52, Q58]

Keywords: market-based environmental policy; carbon taxes; aviation sector; double dividend.

1. - Introduction

The issue of environmental regulation of the aviation sector is much under discussion. The European Parliament has adopted, in July 2006, a report that proposes that "the EU takes a leadership

---

*<anil.markandya@bc3research.org>; <elena.ricci@feem.it>. The authors thank the anonymous referees of *Rivista di Politica Economica* and the editors of the special issue for the useful comments and discussions. The ideas expressed in this paper are those of the authors and not necessarily those of the institutions they belong to.

position in global aviation in order to reduce the climate change impact of aviation" (EU Parliament, 2006). From a policy point of view a carbon tax on aviation is interesting because it is set in the context of climate change, an environmental issue whose interest among public is rapidly increasing. From an economic and methodological point of view it is interesting to study such a tax because it is one of the most complex environmental taxes to design. This is mainly due to the fact that planes are non fixed point-sources and can travel across political and regulatory boundaries.

Aviation is a fast-growing industry and its emissions are currently cause of concerns among experts (IPCC, 1999; Pearce and Pearce, 2000), mainly because this sector is expanding faster than technological improvement in carbon efficiency, so its impacts on emissions are in steady growth (Sewill, 2003). Air travel is the source of many externalities that relate to various pollutants – like NO_x, SO_2, HC, PM_{10}, contrails – noise, congestion and CO_2 emissions. This multi-pollutant nature makes it difficult to set up a tax that takes into account all of the external costs induced on society. More complications arise because its trans-boundary nature makes certain types of taxation policies ineffective, unless they are part of an international agreement. International environmental agreements (IEA) are also needed for issues of international competition (Keen and Strand, 2007).

Many arguments in favour of an aviation tax derive from the fact that this sector is under-taxed. Aviation has always been viewed by governments as a strategic industry and therefore protected (Sewill, 2003; Keen and Strand, 2007; Dings *et* al., 2002). Indeed, currently there are fewer taxes on aviation compared to other productive sectors: in many industrialised countries there are no taxes on aviation fuel, no (or little) VAT, landing fees and airport charges are lower than the market level and there are tax-breaks for passengers through duty-free goods (Sewill, 2003). In Italy, the current tax structure is only aimed at the noise externality[1] and not at the impact of aviation on increasing the concentration of CO_2 in the atmosphere (see Table 1).

---

[1] More precisely, the tax was introduced in 1990 as a national tax and is now

TAB. 1

EXISTING AVIATION CHARGES IN ITALY

ITALY	domestic flights	international flights
VAT (%)	10	–
Airport charges (€)	4.8	4.8
Trip charges (€)	6-9	6-9
Noise tax	See[*]	See[*]

Source: ENAC (2006); OECD (2007).
* See note 1, page 172.

For what concerns CO_2 emissions, the aviation sector is particularly interesting because it accounts for 4% of current anthropic emissions and is the fastest growing source[2] of emission, but – until January 2009 – is was not included in the EU Emission Trading Scheme (EU ETS). International aviation is also excluded from the Kyoto Protocol.

Hence, in the literature there are two main perspectives in favour of an aviation carbon tax (Keen and Strand, 2007). The environmental argument is that the market is capable only to charge users for the benefits of air travel, but not for its negative impacts, thus producing economic inefficiencies (Dings *et* al., 2002). In short, aviation should internalise the external costs it imposes on society. Moreover, being currently under-taxed this sector should be able to sustain an increase in fiscal pressure. The second perspective is that aviation should contribute to public finances as other sectors, as otherwise an additional externality of reduced public services is induced. From the literature, it is known that economic welfare is maximised when all industries pay the

---

a regional tax. The tax base is divided in three categories that pay a different tax rate per ton, above the first 25 ton, for each landing and take-off: 0.33 maximum for 1st category; 0.24 maximum for 2nd category and 0.08 maximum for 3rd category. Exemptions pertain state authorities, health and emergency services and military forces. 100% of the revenue goes to regional environmental authorities to subsidise and compensate municipalities and citizens near the airports (OECD, 2007).

[2] From 1992 the average rate of increase in EU CO_2 emissions from aviation has been on average 5.2% (EUROPEAN COMMISSION, 2004).

same rate of tax (Sewill, 2003) and in the aviation case this is true for corporation, income and labour taxes, but not for VAT and fuel taxes (Sewill, 2003).

In addition, as with all "green-taxes", a tax on aviation CO_2 emissions is able to generate not only a behavioural change that targets the environmental externality, but also revenue for the state, that may be used to achieve other, non necessarily environmental, objectives, *i.e.*, what in the literature is referred to as "the double dividend" (Goulder, 1994).

In this perspective, particularly interesting are the cases in which the effects of green taxes are achieved in a revenue-neutral way, that is without increasing global tax pressure. This can be obtained by using the revenue collected to reduce other pre-existing taxes and, consequently, producing macroeconomic benefits (Markandya, 2006), due to the distortionary effects that non-Pigouvian taxes can have on the market. Literature focuses on two main "second dividends": employment and gross-welfare dividends (Robertson, 2006). The former refers to an increase in employment consequent to the reduction of labour taxes, the latter to the diminished distortions on consumer choice that can be achieved by lowering sales and other taxes (Markandya, 2006). Clearly this concept has a strong appeal as it jointly tackles two crucial policy-goals: better quality environment (first dividend) and economic objectives (second dividend), that are usually part of a trade-off (Carraro and Siniscalco, 1996); however many studies have highlighted how the situation is more complex and that there are many complicated issues that can undermine these effects (Markandya, 2006; Markandya, this issue).

Although the effectiveness of green taxation on such major issues is still quite uncertain and discussed, it is important to underline the potential of the concept of "tax shifting" included in the double-dividend approach. Some economists believe in a progressive shift from taxation of "goods", like production and labour, to the taxation of "bads", like pollution or environmental damage, in order to favour and support "virtuous behaviour" (Pearce, 1991; Repetto and Schwartz, 2000). This is intended to be a strategy for the future that can move towards a more advanced management of sustainable economics. Green taxes can represent a first step towards a new taxation system that focuses not only on the amount of profits gained, but also on the way in which these profits are achieved, socially and environmentally.

It is also important not to forget the appeal that a fiscally neutral eco-tax reform may have on public acceptance and the cultural changes, in terms of diffusion of environmental awareness (Green Taxes "extra dividends").

Because of the unsolved debate in the literature and because of the complex management of the many possible uses of the revenue derived from environmental taxes and their effects, there is, in our opinion, the need for instruments and methodologies that can help the multi-dimensional comparison among all available options, including their mix. Therefore, in the last part of this article, the proposal of a Green Taxation Matrix (GTM) – to help the complex integration of all the different possible effects (*i.e.*, on environment, employment, gross-welfare, R&D, society and culture) of each green tax – is put forward.

This matrix is intended as an instrument that can construct a cognitive map for policy makers to deal with all the different options of revenue management and help them in their decisions. In this work, a first draft of this matrix, that takes into account the options that are currently discussed in the literature and other ones that we consider of interest, is proposed. This model can be enlarged to include new proposals. For each option, a simple index of the potential performance of the tax revenue is designed; these indices can be refined further and modified *ad hoc* to reflect the peculiar characteristic of each case-study.

2. - Designing the Tax

From theory we know that the level of Pigouvian taxes should be set by comparing marginal damage costs with marginal abatement costs. In the literature, aviation marginal damage costs have been estimated; though particularly for those regarding CO_2, there is still no general consensus due to technical difficulties in estimating future damages of a complex dynamic process and to differences in ethical principles that lead to different modelling decisions, like for example discount rates. Pearce and Pearce (2000) estimate the unit shadow prices for CO_2 damage induced by air travel to be 0.029 £/kg. Dings *et* al. (2002) estimate that external costs are approximately 5% of the average ticket price for a 6,000 Km flight and about 20-30% for a 200 Km journey. More specifically, the European Environmental Agency (EEA) has set total external costs of aviation for Italy at 218 € for 1,000 tonne-km and 52 € for 1,000 passenger-km (EEA, 2001).

Marginal abatement costs are sensitive information to which it is difficult to have access. Therefore it has not been possible to conduct the study for the "optimal" tax. Indeed because of a lack of such information it often happens that policy is set politically rather than economically (Pearce and Pearce, 2000).

POSSIBLE STRUCTURES OF THE TAX. Air travel is a multi-pollutant activity and the amount and timing of the externalities that derive from it vary depending on the nature of these "pollutants". Because of the difficulty in designing one instrument to target many differently operating externalities, we focus on only one damaging effect: that of increasing anthropic CO_2 emissions. The literature itself recommends to target only one externality for each instrument (Lanza and Sanmarco, 1996). In any case an extension to cover other externalities would involve adding up the taxes appropriate for each externality (Pearce and Pearce, 2000).

In the literature there are three main types of aviation taxes under consideration:

– *Excise tax on fuel.* CO_2 emissions depend on the amount (and type) of fuel burned, so one option is to put a tax on fuel. This tax targets the externality quite well as it acts on both efficiency and level of activity. The only incentive it could lack is that of favouring investments in the production of less carbon-intensive fuels, but this could be easily changed once more types of fuel were to become available. This tax would have very low administrative costs as it would be integrated with activities that take place regularly for commercial reasons (Smith, 1995), but it would suffer "leakage" problems (*i.e.*, companies would seek to buy fuel in the lowest taxed areas). In fact, such a tax needs to be implemented at least at an EU-level[3], in order not to penalise the competitiveness of single state industries, but, even in this way, it would need restrictions on refuelling. This is not an easy accomplishment because of the existence of very strict international rules on taxing aviation activities. There are also many Aviation Service Agreements (ASAs) among states that would complicate the implementation of a fuel tax on international flights (Pearce and Pearce, 2000). For long-haul flights, although technically feasible, there would be limited leakage potential due to economic reasons: tanking abroad would require the need of larger fuel tanks or inconvenient redirections of flights and the extra weight would increase fuel usage considerably, reducing the net savings (ECON, 2005). When introducing IEAs, problems could arise when deciding how to spend or divide the revenue among states. Some countries have implemented such a tax alone, but at very low levels. If tax levels are low, potential net savings from fuelling abroad are small, considering the additional distance and weight (ECON, 2005); but this also depends on the geo-political location of the state.

– *Ticket tax*. This is an *ad valorem* charge on the sales of tickets (cargo and/or passenger). This though is not a true environmental tax because it is based on industries' revenue and not on their polluting emissions. Such a tax, in fact, imposes different charges on the same amount of externality, proportional to the price charged. VAT charges are of such form (Keen and Strand, 2007).

– *Passenger or trip tax*. Some countries, like the UK, have opted for a passenger tax. This tax is very easily managed and does not violate major international agreements (Wit *et* al., 2002), but it does not target the externality very well as it is not directly related to the distance travelled. Consequently, there are no incen-

---

[3] Before 2003 Single EU Member State implementation would not be permitted by the EU Mineral Oil Directive (92/81/EC). Confirming the raising interest around such taxes, the EU Council Directive 2003/96/EC allows taxation on aviation fuels for domestic (and with bilateral agreements also intra-Community flights).

tives on consumers to chose shorter flights (unless more levels of differentiation of the rate are introduced). Furthermore, as the tax depends only on the number of flying passengers, there are also no incentives for airlines to reduce the number of flights with empty seats and no incentives for producers/buyers of engines, airplanes or fuel to increase the "CO_2 efficiency". The other possibility is a seat-tax which, although still not closely linked with emissions, would at least create the incentives to fly "fuller" planes. Passenger or seat taxes do not affect pure freight flights, unless purposely integrated with other *ad hoc* instruments. Nevertheless, these kinds of taxes still increase the price of each flight for consumers, so they do have, as all other alternatives, some incentive on consumers to reduce the number – and distance, if appropriately differentiated – of flights purchased. Though, the Norwegian experience shows that these kinds of taxes are more unpopular than other ones (ECON, 2005).

These are the options that are mostly discussed in the literature and in real-life applications, because of their low-cost management. The next set of taxes has a completely different approach and is based on a direct tax. One of their distinctive characteristics is the possibility for a country of imposing a tax only on the emissions produced over its national territory.

– *Direct tax*. A first-best solution could be that of a direct tax on emissions, this could be achieved by putting a CO_2 sensor on each plane or a fuel registry to record the exact amount of fuel used. This kind of tax perfectly addresses the externality and accurately spreads the burden among polluters creating the correct incentives to reduce emissions (targeting both efficiency and amount of activity), but it is costly to implement (Smith, 1995).

– *Distance-efficiency tax*. The above tax structure could be simplified by estimating the emissions of each plane instead of measuring them directly. This requires manufactures to define the "CO_2 efficiency" of each type of plane (engine + body) in ton CO_2/litre or ton CO_2/km. As the aviation sector is strictly monitored, it would be easy to calculate the amount of km each plane flies over the Italian territory. To simplify further, an average route kilometer could be used. Such an estimate of the externality induced by each flight is not exact but it is a good approximation; in particular, what is lost is the impact of flight conditions on emissions. This tax would create the same incentives as the direct one, but with much lower administrative costs.

– *Distance-weight tax.* Another level of approximation would be to calculate an average "weight-CO_2 efficiency" and impose a tax on ton $CO_2/(km*ton$ plane), as flight emissions mainly depend on distance and plane weight. This type of tax still creates the incentives on efficiency of plane design and amount of activity with even smaller management difficulties. This kind of tax, in Italy, would have low administrative costs as it could be related to the pre-existing tax[4] on aviation noise that is based on the weight of planes.

ECON (2003) shows how efficiency incentives are important, as there is scope for improvement both in the short and long run. With such improvements, however, there is the possibility of a "rebound effect" under which the gains in efficiency would result in lower cost and prices, thus encouraging more travel.

THE LEVEL OF THE TAX. The difficulty in calculating the "optimal-Pigouvian" level of the tax leaves large scope for policy-makers to decide on the tax to impose. Literature strongly suggests to introduce low levels of tax, that can be gradually increased over the years in order to minimise adjustment costs (INFRAS, 1996; Andersen, 1994). Many countries do indeed impose low levels of taxes also for competitiveness reasons. This does, however, hinder the environmental effectiveness of the tax (ECON, 2005). Indeed, Weitzsacker and Jesinghaus (1992) and Andersen (1994) state that only optimal taxes[5] have the power to induce the required changes.

A powerful option is that of differentiating the tax level to reflect fuel use and emissions. Although this is not ideal from an environmental point of view, as all CO_2 emissions have the same damage potential, it is instead interesting from an economic perspective, because it can help reduce some unwanted effects. All of

---

[4] The advantages of linking Green Taxes to existing tax structures are discussed in SMITH S. (1995).

[5] That reflect environmental damage costs.

the previous tax structures can be differentiated. Passenger taxes are usually differentiated with respect to national and international destination which approximates, in a very rough way, distance (as in the case of the UK and Norwegian taxes). Fuel taxes could be differentiated if more eco-friendly fuels were introduced. The last three taxes could also be differentiated by weight, distance and, more interestingly, kind of flight: passenger or cargo. The existing Italian noise tax is, in fact, differentiated with respect to weight classes. Differentiation should penalise short hauls that have more substitution possibilities. The environmental effects of such differentiation depend, however, on the consequent change in the demand for other transport services and their carbon intensity.

DEMAND ELASTICITY. The price elasticity of demand for any mode of travel is related to the possibilities of substitution. For the particular case of air travel, on certain routes, various types of substitution can be identified (Brons *et* al., 2001): *i)* intra-modal substitutions, *ii)* inter-modal substitutions, *iii)* substitutions of destination, *iv)* substitution with non-transport goods. The first refers to the competition between different airlines, while the second to other means of transport, whose success is determined by geographic, economic and demographic factors (Brons *et* al., 2001). In this particular market, if prices increase there is also the possibility of choosing different destinations with similar attributes, or even completely different goods that yield the same utility (third and fourth case). Especially for the last two types of substitution, the level of substitutability strongly depends on the nature of travel. Consumers can be divided into two main groups that have major differences: business and leisure passengers. The latter aim at maximising their utility/satisfaction derived from the trip subject to a budget constraint, while business travellers try to minimise the costs of achieving a certain level of output, for which travel is an input (Brons *et* al., 2001). These two kinds of consumers are likely to respond differently to a raise in price due to the introduction of a tax on air travel. In particular there are more goods that compete with leisure travel, so this kind of passenger is more price sensitive. In short, the elasticity of substitution varies with distance to be travelled, location of departure, reason of travel, but also with income, as we consider next.

DISTRIBUTIONAL EFFECTS. The tax structures discussed above are not related to income, so if the demand for air travel was homogeneous among all consumers these taxes would probably be regressive, in the sense that they would affect a bigger proportion of the income of poorer people. In the literature, however, air travel is considered a luxury good and its demand is higher among high-income individuals (Sewill, 2003; Brons *et* al., 2001). This means that a tax on air transport would mostly affect high-income consumers, because their budget share for air travel is bigger in comparison to low-income individuals. Negative distributional effects can, in any case, be limited by gradually increasing the tax over time (INFRAS, ECOPLAN, 1996).

3. - The Model

Our aim is to propose a model of evaluation of the economic effects of the introduction of an aviation carbon tax in Italy, although the methodology is more general and could be applied to other green taxes and to other countries. In the literature, many different types of models – at different levels of complexity – are available (partial-equilibrium models, input-output models, general equilibrium models and macro-economic models).

The aim of constructing a model that can be used by policy-makers has lead us to organise the analysis in a way that it can be easily applied in concrete situations. Therefore, in this work, we propose an integrated three-step analysis based on:

1. a partial equilibrium model to estimate demand and supply functions and evaluate the effects of the introduction of the environmental tax on market equilibrium, industry revenue, targeted sector labour market, dead weight loss, consumer and producer surplus and state revenue;

2. an Input-Output model to take into consideration the inter-sectoral relationships and evaluate the effects on all the sectors of the economy due to changes in the final demand of the sector targeted by the tax;

3. a Green Taxation Matrix (GTM) that is our proposal of an instrument that can compare the multiple possibilities of tax revenue management, and evaluate the relevance of the double dividend.

3.1 *Partial Equilibrium Model*

ESTIMATION OF CURRENT DEMAND AND SUPPLY CURVES FOR AIR TRAVEL IN ITALY

As a starting point of our analysis, we use a partial-equilibrium model instead of more sophisticated simulation models. The advantage of this model is that it is simple and easily applicable to real cases, as it only requires the knowledge of information that should be available without difficulty to policy-makers. It is also very transparent as there are no hidden assumptions and its limitations are easily found in the literature. This means that the results and their relevance are clearly interpretable. Obviously, using a simple model has also drawbacks and the partial equilibrium model, in particular, isolates the market under discussion from the rest of the economy, that is considered to be static (*ceteris paribus* assumption). The main additional restrictive assumption introduced in this analysis is the assumption of linear demand and supply functions. The real functions will rarely have a linear behaviour, but for small movements from the equilibrium any function can be approximated by a linear one.

Operatively, we start from current equilibrium values of price, quantity, supply and demand elasticities (P^*, Q^*, e_S^* and e_D^*) and estimate the corresponding demand and supply functions, *i.e.*, α, β, γ and δ.

(1)
$$P_S = \alpha + \beta Q_S + t$$

(2) $$P_D = \gamma + \delta Q_D$$

When a tax on the good is introduced the supply function changes: a term t – that represents the tax that needs to be paid for each unit of good – is added[6]. Therefore, a new equilibrium arises, denoted by new levels of price and quantity (P^{**}, Q^{**}).

Current price elasticity of demand and supply. The price elasticity of demand for air travel has been estimated in a study by the Tinbergen Institute (Brons et al., 2001) that conducted a meta-analysis on 37 international studies[7] that estimate price elasticities for passenger air transport. The distribution of such estimates shows a bi-modal structure that is associated to the different estimates that arise from case-studies that consider leisure or business passengers (Brons et al., 2001).

Due to the impossibility of finding data regarding the expenditure of firms in air travel, our analysis takes into consideration the effects of the tax on the total market and on leisure travel population only (Table 2).

TABLE 2

PRICE ELASTICITY OF DEMAND VALUES FOR AIR TRAVEL

$e_D{}^*$	mean	standard deviation
Total market	-1.146	0.619
Leisure travel	-1.267	0.624

Source: Adapted from Brons M. et al. (2001).

---

[6] The choice of such a structure is due to the fact that this is a tax on quantity of emission generated and not on the revenue of the industry, as the polluting effects are more related to quantity of output than profits. Assuming that preferences and production strategies remain the same, we use the previously estimated parameters α, β, γ and δ to find the new equilibrium point.

[7] The set included a total of 204 observations. The case-studies analysed refer mainly to developed countries, mostly Europe, USA and Australia.

For each case, three values of demand elasticity have been used: mean value, mean value plus one standard deviation and mean value minus one standard deviation. For brevity, in the following sections, only mean value cases will be reported; the other cases are available from the authors on request.

No studies regarding the price elasticity of supply of the aviation sector were found in the literature. Therefore, we tested different values ranging between 0.5 and 1.5 (*i.e.*, covering both inelastic and elastic supply functions). The results are quite robust to changes in the value of elasticity within the evaluated range, therefore, we here report only the central case where the supply is unit elastic. The results for the other cases are available from the authors on request.

Current revenue of air travel sector. To calculate P^* and Q^* we need information on current industry revenue. The following data will be analysed:

– Domestic flights: where the tax is imposed on all routes within the country;

– European flights: where the tax is imposed on all flights departing and arriving to Italy to and from continental Europe (EU + non-EU countries);

– Italian vectors: where only flights from Italian airlines[8] on all national and international routes are considered;

– Leisure travel: where we study the effects of the tax on all flights bought by Italian households.

Data are taken from *Annuario Statistico 2005* (ENAC, 2006) edited by ENAC (Ente Nazionale per l'Aviazione Civile), *Conti Economici Nazionali*-ISTAT (ISTAT, 2007*b*) and *Government Finance Statistics from International Monetary Fund* (IMF, 2007) and refers to 2005. Data for the analysis of leisure travel demand are taken from *I consumi delle famiglie* (ISTAT, 2007*a*), and refers to the annual expenditure in air travel of Italian families in 2005 (Table 3).

---

[8] Namely: Air Dolomiti S.p.A., Air Italy, Air One S.p.A., Air Vallee S.p.A., Alitalia S.p.A., Alpi Eagles S.p.A., Blue Panorama, Club Air, Eurofly S.p.A., Itali Airlines, Lauda Air, Livingston S.p.A., Meridiana, My Way, Neos S.p.A., Windjet S.p.A.

TABLE 3

CURRENT REVENUE OF THE AVIATION SECTOR

market	total revenue [€]
Domestic flights	3,154,671,997
Continental European flights	9,884,947,722
Italian carriers	6,691,728,478
Leisure travel	1,898,645,075

Source: ENAC (2006); ISTAT (2007*a*).

We were able to find data regarding the aviation sector labour market only for Italian-based carriers (Table 4), therefore the analysis relative to the possible effects of tax on labour in the aviation sector is conducted only for these carriers.

TABLE 4

AVIATION SECTOR EMPLOYMENT DATA

Total Labour Costs (L^*)	1,117,421,630 €
Average Salary (W^*)	62,506 €
Employees (E)	17,877
Ratio between labour costs and revenue (L^*/R^*)	0.166986

Source: ISTAT (2007*b*).

Without loss of generality, the initial equilibrium price (P^*) can be set to 1. Consequently the initial equilibrium quantity (Q^*) is numerically equal to current industry revenue. This means that, in our analysis, we will use as quantity unit the quantity of good that today has a market price of 1€. Although it is less straight forward to link this quantity unit to its environmental damage/externality – because it does not correspond to any "physical" unit – using $P^* = 1$ has the advantage of *(i)* being able to work implicitly in terms of $[t/(P^*)]$, that is useful for welfare considerations and of *(ii)* not linking results to the chosen quantity unit – therefore all five structures of tax can be analysed[9]. It must be

---

[9] Simulations of fuel taxes can be run by knowing the percentage of fuel costs on total costs, that for european carriers is estimated to be 20% (AEA, 2006).

stressed that this is a tax on quantity and not on revenue, although it can be expressed as a fraction of present price P^*.

TABLE 5

CURRENT MARKET EQUILIBRIUM VALUES

	P^*	Q^*
Domestic flights	1	3,154,671,997
Continental European flights	1	9,884,947,722
Italian carriers	1	6,691,728,478
Leisure travel	1	1,898,645,075

Source: Adapted from ENAC (2006); ISTAT (2007a).

Given all previous data and assumptions, the estimates of the demand and supply functions for the four different markets used in this analysis are:

(3) Domestic flights:
$$\begin{cases} P_S = 3.170 \cdot 10^{-10} Q_S + t \\ P_D = 1.872 - 2.766 \cdot 10^{-10} Q_D \end{cases}$$

(4) European flights:
$$\begin{cases} P_S = 1.012 \cdot 10^{-10} Q_S + t \\ P_D = 1.872 - 0.883 \cdot 10^{-10} Q_D \end{cases}$$

(5) Italian carriers:
$$\begin{cases} P_S = 1.494 \cdot 10^{-10} Q_S + t \\ P_D = 1.872 - 1.304 \cdot 10^{-10} Q_D \end{cases}$$

(6) Leisure travel:
$$\begin{cases} P_S = 5.267 \cdot 10^{-10} Q_S + t \\ P_D = 1.789 - 4.157 \cdot 10^{-10} Q_D \end{cases}$$

ESTIMATION OF THE EFFECTS OF THE TAX

In the following analysis we consider three particular values of the tax that correspond to an increase of 5%, 10% and 15%, respectively, of the current price P^*.

TAB. 6

EFFECTS OF THE TAX ON MARKET EQUILIBRIUM

	t_{max}	P^{**}	Q^{**}	Q/Q^*	P	R	R/R^*
Domestic flights							
$t = 0.05$	1.87	1.023	$3.070 \cdot 10^9$	-0.027	0.023	$-12.694 \cdot 10^6$	-0.004
$t = 0.10$	1.87	1.047	$2.986 \cdot 10^9$	-0.053	0.047	$-29.312 \cdot 10^6$	-0.009
$t = 0.15$	1.87	1.070	$2.902 \cdot 10^9$	-0.080	0.070	$-49.856 \cdot 10^6$	-0.016
European flights							
$t = 0.05$	1.87	1.023	$9.621 \cdot 10^9$	-0.027	0.023	$-39.775 \cdot 10^6$	-0.004
$t = 0.10$	1.87	1.047	$9.357 \cdot 10^9$	-0.053	0.047	$-91.849 \cdot 10^6$	-0.009
$t = 0.15$	1.87	1.070	$9.093 \cdot 10^9$	-0.080	0.070	$-156.222 \cdot 10^6$	-0.016
Italian vectors							
$t = 0.05$	1.87	1.023	$6.513 \cdot 10^9$	-0.027	0.023	$-26.926 \cdot 10^6$	-0.004
$t = 0.10$	1.87	1.047	$6.334 \cdot 10^9$	-0.053	0.047	$-62.178 \cdot 10^6$	-0.009
$t = 0.15$	1.87	1.070	$6.156 \cdot 10^9$	-0.080	0.070	$-105.756 \cdot 10^6$	-0.016
Leisure travel							
$t = 0.05$	1,79	1.022	$1.846 \cdot 10^9$	-0.028	0.022	$-12.351 \cdot 10^6$	-0.007
$t = 0.10$	1,79	1.044	$1.792 \cdot 10^9$	-0.056	0.044	$-27.042 \cdot 10^6$	-0.014
$t = 0.15$	1,79	1.066	$1.739 \cdot 10^9$	-0.084	0.066	$-44.074 \cdot 10^6$	-0.023

Source: Own calculations.

All of the results reported in the Tables, except quantities and percentages, are expressed in euros and relate to the four types of data on aviation previously discussed[10].

Table 6 shows the effects of the three different values of tax on the market equilibrium. Taxes that correspond to 5%, 10%, 15% of the current price, impose an increase in price – for the first three markets examined – in the order of, respectively, 2.3%, 4.7%, 7.0%, as part of the tax is borne by the producers. For the same taxes, the quantity is reduced by 2.7%, 5.3%, 8.0%. These percentage values change only slightly in the case of leisure travel of Italian families.

This change in quantity is the actual "first dividend" of the tax and corresponds to a reduction in the polluting activity. In this specific case, it is difficult to calculate the impact in physical units (tonCO$_2$ avoided), as in the aviation market the relation between

---

[10] All analyses have been performed with the statistical program R (R DEVELOPMENT CORE TEAM, 2007).

the unit of good sold/purchased and its relative emissions is not unequivocal since it is dependent from many other factors[11] that do not relate to the environment.

Table 6 also reports the theoretical maximum value of the tax (t_{max}), which is the value of the tax for which production goes to zero, that is a useful benchmark for comparing the value of the tax considered[12].

In the same Table, the effects of the three values of tax on the revenue of the aviation sector (R) are reported. All of the four cases described in the Table have price elasticity of demand smaller than -1; consequently, the change in revenue of the sector (ΔR) after the tax is always negative. The industry revenue, in the general case with elasticity of demand equal to -1.146 and supply elasticity equal to 1, with a tax that corresponds to 5%, 10% and 15% of the current price, is reduced by only 0.4%, 0.9% and 1.6% respectively, and these values increase with the increase[13] in demand elasticity. From these results, the effects of the tax on the market seem to be acceptable. This is due to the fact that the levels of the tax chosen are, purposely, quite low to reduce disruption in the market and opposition to the tax.

TABLE 7

EFFECTS OF THE TAX ON THE AVIATION
INDUSTRY EMPLOYMENT

	labour expenditure[14]	E	E/E^*
Italian vectors			
$t = 0.05$	$-4.496 \cdot 10^6$	-72	-0.004
$t = 0.10$	$-10.383 \cdot 10^6$	-166	-0.009
$t = 0.15$	$-17.660 \cdot 10^6$	-283	-0.016

Source: Own calculations.

---

[11] For example: airline, ticket class, time of flight, advance booking, importance of the route/airport, fullness of the plane, airline airport agreements (in addition to distance and added weight).

[12] Because of the linear assumptions, this maximum tax level of 187% is not necessarily realistic. Our model is more accurate the smaller the tax is, compared to t_{max}.

[13] In absolute values.

[14] With elasticity of substitution in production equal to one (KEEN M., STRAND J., 2007).

Table 7 reports the effects of the tax on employment of the aviation sector. Results show the number and the percentage of employees that would be dismissed (ΔE and $\Delta E/E^*$) under the two strong assumptions: *(i)* that the input mix of production remains the same and *(ii)* that the labour input can be varied by the producer in the short-run. While the first assumption is acceptable and coherent with the model, the second assumption is unlikely to take place in practice, so we can view these values as the maximum effect that the tax could have on labour. This is even more so in the Italian labour market, that presents stronger rigidness than others. These values are to be considered as maximum effects also because – if the revenue of the state is devolved to reduce labour taxes – green taxes may have positive effects on global employment, thus also in the aviation sector.

Table 8 shows the effects of the tax on the revenue of the state. $t_{max}/2$ is the value of the tax for which the state revenue reaches its maximum value, equal to *max* R_{state}. R_{state}/R^*, instead, indicates the fraction of current industry revenue (R^*) that would go to the state because of the tax.

TABLE 8

EFFECTS OF THE TAX ON STATE REVENUE

	t_{max}	$t_{max}/2$	R_{state}	*max* R_{state}	R_{state}/R^*
Domestic flights					
$t = 0.05$	1.87	0.94	$0.153 \cdot 10^9$	$1.477 \cdot 10^9$	0.049
$t = 0.10$	1.87	0.94	$0.299 \cdot 10^9$	$1.477 \cdot 10^9$	0.095
$t = 0.15$	1.87	0.94	$0.435 \cdot 10^9$	$1.477 \cdot 10^9$	0.140
European flights					
$t = 0.05$	1.87	0.94	$0.481 \cdot 10^9$	$4.628 \cdot 10^9$	0.049
$t = 0.10$	1.87	0.94	$0.936 \cdot 10^9$	$4.628 \cdot 10^9$	0.095
$t = 0.15$	1.87	0.94	$1.364 \cdot 10^9$	$4.628 \cdot 10^9$	0.140
Italian vectors					
$t = 0.05$	1.87	0.94	$0.326 \cdot 10^9$	$3.132 \cdot 10^9$	0.049
$t = 0.10$	1.87	0.94	$0.633 \cdot 10^9$	$3.132 \cdot 10^9$	0.095
$t = 0.15$	1.87	0.94	$0.923 \cdot 10^9$	$3.132 \cdot 10^9$	0.140
Leisure travel					
$t = 0.05$	1.79	0.89	$0.092 \cdot 10^9$	$0.849 \cdot 10^9$	0.049
$t = 0.10$	1.79	0.89	$0.179 \cdot 10^9$	$0.849 \cdot 10^9$	0.096
$t = 0.15$	1.79	0.89	$0.261 \cdot 10^9$	$0.849 \cdot 10^9$	0.141

Source: Own calculations.

Table 9 shows the welfare effects of the tax, here shown in terms of the change in industry revenue (ΔR), State revenue (ΔR_{state}), Consumer Surplus (ΔCS), Producer Surplus (ΔPS) and Dead Weight Loss (DWL), calculated in a traditional manner. In standard economics, the DWL is considered as the social cost of the tax (Varian, 1990): as no-one (Consumer, Producer or State) gains from the relative reduction in output it is considered a pure loss. Though, in the case of green taxes this needs to be interpreted in a different way: the DWL still represents a loss of the value of the lost transactions, though – if the tax is calculated correctly – this willingness to trade is motivated by considerations that do not fully take into account the social/environmental costs of those transactions. Therefore, the DWL described in this analysis is to be considered an indicator of the transaction loss induced by the tax, bearing in mind that the tax purposely reduces the air travel activity because the expected value of the damage associated is higher than the value of the transactions.

For all values of the tax and of the price elasticities considered in this analysis, the DWL is positive (*i.e.*, there is a loss, albeit small). We also note that the change in the revenue of the aviation sector (ΔR) is much smaller than the revenue that is collected by the state, that is instead equal, by definition, in absolute terms, to $PS + CS - DWL$. The whole analysis shows that DWL increases with the increase of supply and demand price elasticity[15]. Consumer surplus decreases as $e_S{}^*$ increases and as $e_D{}^*$ decreases[15] while producer surplus increases as $e_S{}^*$ increases and as $e_D{}^*$ decreases[15].

The fraction of tax paid by consumers – that corresponds to 0.47 in the first three cases and to 0.44 in the leisure travel case – decreases with $e_D{}^*$[15], as it would be expected. This is interesting for comparing effects on populations with different elasticities, but not for the same group as policy-makers are unable to influence price elasticities, except in the long run through policies aimed at changing consumer preferences or producers input mix.

As already mentioned, the results are stable with respect to changes in the value of the price elasticity of supply. In particu-

---

[15] See note 13.

TABLE 9

WELFARE EFFECTS OF THE TAX

	ΔR	R_{state}	CS	PS	DWL
Domestic flights					
$t = 0.05$	$-12.69 \cdot 10^6$	$153.52 \cdot 10^6$	$-72.52 \cdot 10^6$	$-83.11 \cdot 10^6$	$2.11 \cdot 10^6$
$t = 0.10$	$-29.31 \cdot 10^6$	$298.62 \cdot 10^6$	$-143.08 \cdot 10^6$	$-163.97 \cdot 10^6$	$8.42 \cdot 10^6$
$t = 0.15$	$-49.86 \cdot 10^6$	$435.30 \cdot 10^6$	$-211.67 \cdot 10^6$	$-242.58 \cdot 10^6$	$18.95 \cdot 10^6$
European flights					
$t = 0.05$	$-39.77 \cdot 10^6$	$481.05 \cdot 10^6$	$-227.24 \cdot 10^6$	$-260.41 \cdot 10^6$	$6.60 \cdot 10^6$
$t = 0.10$	$-91.85 \cdot 10^6$	$935.71 \cdot 10^6$	$-448.32 \cdot 10^6$	$-513.78 \cdot 10^6$	$26.40 \cdot 10^6$
$t = 0.15$	$-156.22 \cdot 10^6$	$1{,}363.97 \cdot 10^6$	$-663.26 \cdot 10^6$	$-760.10 \cdot 10^6$	$59.39 \cdot 10^6$
Italian vectors					
$t = 0.05$	$-26.93 \cdot 10^6$	$325.65 \cdot 10^6$	$-153.83 \cdot 10^6$	$-176.29 \cdot 10^6$	$4.47 \cdot 10^6$
$t = 0.10$	$-62.18 \cdot 10^6$	$633.44 \cdot 10^6$	$-303.50 \cdot 10^6$	$-347.81 \cdot 10^6$	$17.87 \cdot 10^6$
$t = 0.15$	$-105.76 \cdot 10^6$	$923.36 \cdot 10^6$	$-449.00 \cdot 10^6$	$-514.55 \cdot 10^6$	$40.20 \cdot 10^6$
Leisure travel					
$t = 0.05$	$-12.35 \cdot 10^6$	$92.28 \cdot 10^6$	$-41.29 \cdot 10^6$	$-52.31 \cdot 10^6$	$1.33 \cdot 10^6$
$t = 0.10$	$-27.04 \cdot 10^6$	$179.25 \cdot 10^6$	$-81.41 \cdot 10^6$	$-103.15 \cdot 10^6$	$5.31 \cdot 10^6$
$t = 0.15$	$-44.07 \cdot 10^6$	$260.92 \cdot 10^6$	$-120.36 \cdot 10^6$	$-152.50 \cdot 10^6$	$11.94 \cdot 10^6$

Source: Own calculations.

Note: DWL = R_{state} + CS + PS.

lar, changing e_S within the range (0.5; 1.5), P^{**} and Q^{**}, i.e., the new equilibrium point, do not vary more than one percent with respect to the situation with $e_S = 1$. This implies also stability of the state revenue. The parameter that changes the most is the fraction of tax paid by consumers (or producers). The latter is, indeed, significantly influenced (only) by the *ratio* between the two price elasticities of supply and demand. It can be shown that this fraction, given by $(P^{**} - P^*)/t$, can be expressed as: $(P^{**} - P^*)/t = \delta/(\delta + \beta) = 1/(1 - e_D/e_S)$. In our case, considering all the combinations of values of e_S and e_D, this fraction ranges between 22% and 74%. As expected, the smaller, in absolute terms, is the elasticity of demand with respect to the elasticity of supply, the bigger is the fraction of tax paid by consumers, and *vice versa*.

3.2 *Input-Output Model*

As the economy is a complex network of relationships, the effects of any tax on the final demand of a good are not limited to the good itself, but impact on other goods and services. In order to capture these systemic effects we use an Input-Output (I-O) model that considers the relationships between the various economic sectors and provides a measurement of the monetary flows that occur at the production sector level, thus highlighting the links of interdependence existing among them.

The main limitation of the Input-Output approach is the fact that it assumes the structure of the economy to be fixed, *i.e.*, constant I-O coefficients and elasticity of substitution (INFRAS, ECOPLAN, 1996), while the real-life inter-relations between the sectors are more likely dynamic. This assumption of "steadiness" is coherent with the previous part of the model, that is intended and adequate (INFRAS, ECOPLAN, 1996) only for short or medi-

TABLE 10

TOTAL EFFECTS OF THE TAX ON WHOLE ECONOMY

	R	Total loss of induced expenditure
Domestic flights		
$t = 0.05$	$-12.694 \cdot 10^6$	$-27.312 \cdot 10^6$
$t = 0.10$	$-29.312 \cdot 10^6$	$-63.071 \cdot 10^6$
$t = 0.15$	$-49.856 \cdot 10^6$	$-107.274 \cdot 10^6$
European flights		
$t = 0.05$	$-39.775 \cdot 10^6$	$-85.582 \cdot 10^6$
$t = 0.10$	$-91.849 \cdot 10^6$	$-197.627 \cdot 10^6$
$t = 0.15$	$-156.222 \cdot 10^6$	$-336.136 \cdot 10^6$
Italian vectors		
$t = 0.05$	$-26.926 \cdot 10^6$	$57.936 \cdot 10^6$
$t = 0.10$	$-62.178 \cdot 10^6$	$-133.786 \cdot 10^6$
$t = 0.15$	$-105.756 \cdot 10^6$	$-227.551 \cdot 10^6$
Leisure travel		
$t = 0.05$	$-12.351 \cdot 10^6$	$-26.575 \cdot 10^6$
$t = 0.10$	$-27.042 \cdot 10^6$	$-58.186 \cdot 10^6$
$t = 0.15$	$-44.074 \cdot 10^6$	$-94.833 \cdot 10^6$

Source: Own calculations.

um term analyses, as it is realistic to assume that major structural changes do not arise in short time horizons.

More precisely, we are interested in the variation of induced expenditure in the different sectors, due to the change in demand and expenditure of the aviation sector, caused by the levying of the tax. As demand for air travel decreases, there is a variation in induced demand of other industrial sectors proportional to the amount of intermediate goods used as inputs in the aviation sector. Table 10 reports the aggregate effects on the economy. For example, a decrease of $12 \cdot 10^6$ € in air travel expenditure, due to a 5% tax on domestic flights, creates a loss of induced expenditure of $27 \cdot 10^6$ € on the whole economy. What our model is not able to consider are the effects on the other sectors due to a variation of the purchasing power of consumers subject to the tax.

Graph 1 shows the quantitative relevance of the effects of the tax, through the reduced demand in air travel, on all sectors of the economy. We have plotted the effects of 1€ of expenditure loss, in order to highlight the transformation coefficients. The most affected sectors are those of *Air Transport* itself and *Auxiliary Transport & Travel Agents*[16], *Other Transport, Professional Activities*. Table 11 reports the values of loss of induced expenditure on these sectors.

The results that emerge from this analysis seem to indicate that the effects of such levels of tax are moderate for all sectors of the economy. Therefore, strong opposition to the implementation of this instrument for environmental protection is not justified. An increase in acceptance of the tax, at least from the consumer's side, may arise if the environmental benefit of the tax and the tax shifting concept is stressed. There is, indeed, some evidence in the literature of willingness to pay to compensate the damage caused by own air travel CO_2 emissions (Brower *et al.*, 2008). Moreover, Zero-revenue, hybrid or environment subsidising instruments are more likely to be accepted (Pezzey and Park, 1998). Some survey studies show how an increasing fraction of citizens would be willing to give up fiscal neutrality and pay a certain amount of taxes to reduce the environmental impacts of their actions (Markandya, 2006).

---

[16] In the Tables we refer to it as "Transport Services".

TABLE 11

LOSS OF INDUCED DEMAND FOR
THE MOST INFLUENCED SECTORS

	Air transp.	Transp. serv.	Other transp.	Prof. activities	Other sectors
Domestic flights					
$t = 0.05$	-12,919,234	-2,554,540	-2,241,602	-1,527,857	-19,243,237
$t = 0.10$	-29,833,287	-5,898,982	-5,176,341	-3,528,150	-44,436,769
$t = 0.15$	-50,742,161	-10,033,327	-8,804,216	-6,000,879	-75,580,598
European flights					
$t = 0.05$	-40,481,530	-8,004,476	-7,023,906	-4,787,434	-60,297,358
$t = 0.10$	-93,480,554	-18,484,056	-16,219,707	-11,055,215	-139,239,560
$t = 0.15$	-158,997,071	-31,438,739	-27,587,406	-18,803,342	-236,826,605
Italian vectors					
$t = 0.05$	-27,404,435	-5,418,722	-4,754,913	-3,240,908	-40,818,987
$t = 0.10$	-63,282,731	-12,512,993	-10,980,117	-7,483,954	94,259,813
$t = 0.15$	-107,634,887	-21,282,814	-18,675,610	-12,729,137	-160,322,480
Leisure travel					
$t = 0.05$	-12,570,434	-2,485,571	-2,181,082	-1,486,607	-18,723,698
$t = 0.10$	-27,522,832	-5,442,132	-4,775,456	-3,254,910	-40,995,338
$t = 0.15$	-44,857,193	-8,869,683	-7,783,122	-5,304,910	-66,814,921

Source: Own calculations.

The evaluation of the efficacy of the tax in reducing CO_2 emissions is strongly influenced also by how the revenue collected by the state is re-invested (Carraro and Galeotti, 1995). Indeed, the next section will consider different revenue recycling options, though a complete analysis should include considerations about the CO_2 emission feedbacks of the various strategies. Even if, as already mentioned, it is difficult to evaluate the amount of emission reductions directly induced by the tax through an economic model, it should be noticed that a tax may induce positive secondary effects through other mechanisms. A well designed tax should, in addition to a reduction in the level of activity, induce innovation in production and an increased internalization of the social costs in the purchasing decision process. Moreover, a tax, as all environmental policy instruments, is able to increase interest and awareness, and promote the diffusion of an increased environmental culture. This of public acceptance is an extra dimension that is important to consider as global environmental problems increasingly need the active participation of the public to be tackled.

GRAPH 1

LOSS OF INDUCED EXPENDITURE IN ALL SECTORS OF THE ECONOMY DUE TO 1 LOSS IN AVIATION DEMAND

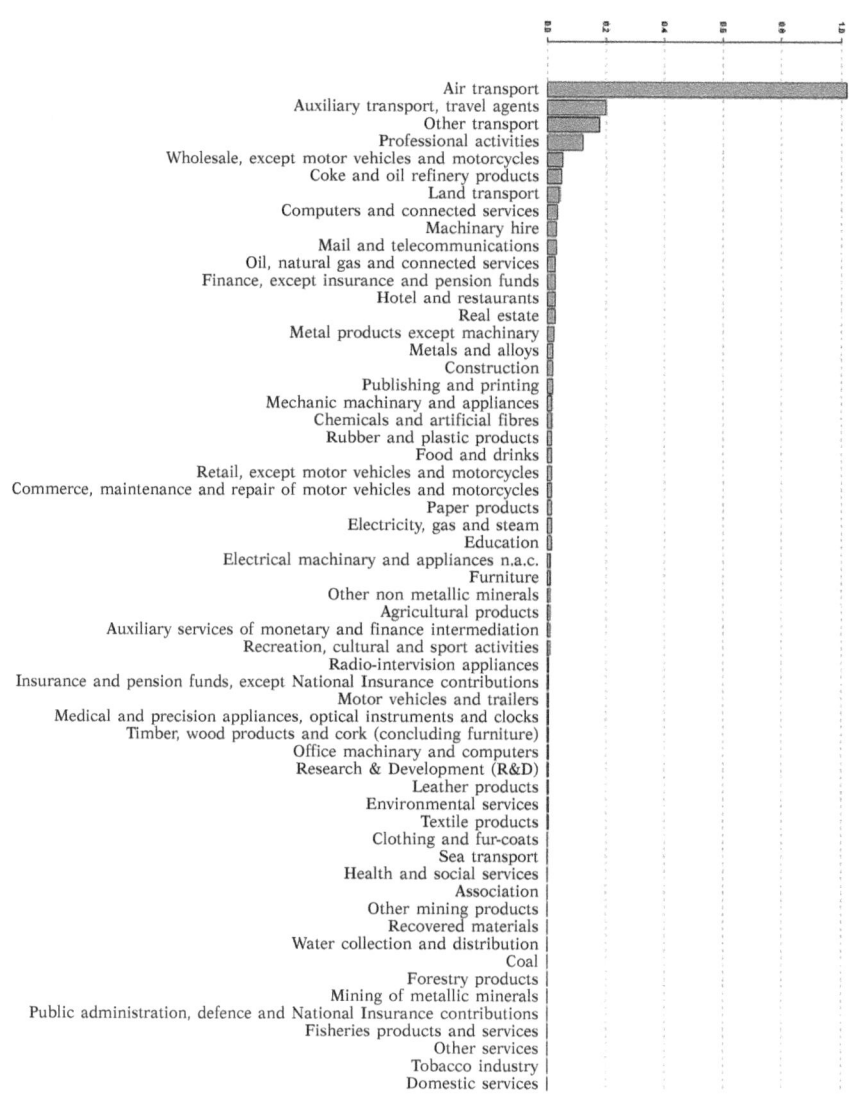

Source: Own calculations.

3.3 *Aviation GT-Matrix*

The third and final part of the paper is aimed at valuing the relevance of the effects of the spending of the state revenue obtained with the tax.

There is still much discussion in the literature about the effects of the double dividend. Its relevance in quantitative terms is crucial to determine the feasibility also of its qualitative potential to promote a structural reform of national taxation systems (Pearce, 1991; Repetto and Schwartz, 2000), that would favour and incentive responsible and virtuous environmental or social behaviour.

Due to the uncertainty and discussion that is still open in the literature, mostly regarding the employment second dividend, we put forward the idea of an instrument capable of comparing various revenue recycling options. The example presented here is very simple, but it can be made more sophisticated if desired.

We suggest integrating the results obtained in the first two steps of the model in a Green Taxation Matrix (GTM) to help the complex integration of all the different potential effects. The aim is again that of designing a simple and neat instrument to help policy-makers in the management decisions regarding the state revenue. More precisely, the object is to develop a matrix with all the likely uses of the state revenue and for each of them formulate an index that evaluates the potential performance of the re-investment of the tax revenue.

As a first proposal of this matrix we simply use as indices the percentage of current expenditure that would be covered by the tax revenue, to give an idea of the extra amount that the state could invest in each specific area[16] or by how much could other taxes be lowered as a result of levying the aviation tax. We also report the percentage of GDP and of total public expenditure for comparison. Each row of the matrix represents a strategy for partitioning the state revenue among the various options, so that all the different possibilities and their mix can be easily compared.

---

[16] If current levels of expenditure are not optimal due to budget constraints.

The case study under examination regards a single tax, therefore a 2D matrix can be developed; this approach can be extended further by overlaying different 2D-GT matrices of various environmental taxes to generate a 3D-GTMatrix able to compare the relevance of various tax revenue uses of different environmental tax options.

Much of the discussion about revenue spending in the literature (Markandya, 2006; Carraro and Siniscalco, 1996) concerns the reduction of labour taxes, leaving global tax pressure unchanged (Bovenberg and de Mooij, 1994; Carraro *et* al., 1996; Bosello *et* al., 2001; Conrad and Löschel, 2005), while earmarking is often criticised (Smith, 1995; Peszko, 1999). The proposed matrix is intended to quantitatively compare and manage all the different options.

For instance, social security contributions can be reduced in general or only for certain categories like, for example, lowest-wage and low skilled workers (INFRAS, ECOPLAN, 1996), environmental-related labour or only for firms that have certain characteristics, like being certified for having a social or environmental responsible behaviour. State revenue could be also used for reducing sales taxes; again this could be done on all goods, and it would in this way be similar to a labour tax, or only on essential or eco-compatible goods. This kind of strategy might also increase the acceptability of the tax by the general public, and is more likely to enhance – rather than hinder – the first dividend of the policy.

Another option of state revenue spending, though not fiscally neutral, could be for example investments on environmental activities – like general protection, prevention or mitigation measures –, victim compensation[17] or sponsoring environmental research and communication projects. For example, the current Italian aircraft noise tax revenue, and that of many other green taxes, is assigned to environmental agencies. Other options could be those of increasing state expenditure in R&D or education and

---

[17] Though compensation measures are always a complicated matter when victims can influence the probability or amount of damage they suffer.

training that are very strategic assets for a country. It is clear that all these examples of revenue spending have the capacity of creating different behavioural incentives that could be used by policy-makers to stimulate changes in society.

In the following Tables we report, for the 5% tax on both domestic and continental European flights, the results concerning five different options of state revenue spending[18], namely: (A) environmental activities, (B) R&D, (C) education and training, (D) employers' National Insurance contributions, (E) employees' National Insurance contributions. The values are calculated comparing the available state revenue with current state expenditure[19] in these sectors. For simplicity of reading, only a few combinations of the possible ones are shown, but potentially the revenue can be spread over all options in any way. In particular, we report the cases where the whole revenue is directed to one option to evaluate its maximum potential effects, and the case where it is equally spread between the five options. It could be of interest to study the case where a percentage of revenue equal to total environmental damage costs is directed to environmental activities and the rest to other uses, like, for example, the reduction of labour taxes. This is an interesting case from an ethical point of view and could be seen as a way to promote weak sustainability, as it implies that damage to one component of the environment can be counteracted by improving some other. In this direction, this part of the revenue should be invested in environmental protection or mitigation activities in general so that the targeted activity could, in some way, internalise its externality. Notice that the revenue should not be spent in existing abatement measures to reduce emissions from the targeted sector, because all the cost-justifiable[20] actions should have already been taken. What might not be included in the marginal abatement schedules that are used to choose the optimal level of pollution are extra sectoral activi-

---

[18] The results for the other levels of tax considered in the rest of the analysis are available on request.

[19] For payroll contributions we do not consider state expenditure, but private contributions induced by state legislation.

[20] For society.

TABLE 12

AVIATION GT-MATRIX FOR A 0.5 TAX AND MEAN VALUES
OF SUPPLY AND DEMAND PRICE ELASTICITIES

		Environ-ment	R&D	Education	Labour Tax employers	Labour Tax employees
Strategy A	% TAX REVENUE	100%	0%	0%	0%	0%
	AMOUNT (10`6 €)	635	0	0	0	0
	% of current expenditure	9,020%	0%	0%	0%	0%
	% total public expenditure	0,092%	0%	0%	/	/
	% GDP	0,045%	0%	0%	0%	0%
Strategy B	% TAX REVENUE	0%	100%	0%	0%	0%
	AMOUNT (10`6 €)	0	635	0	0	0
	% of current expenditure	0%	4,477%	0%	0%	0%
	% total public expenditure	0%	0,092%	0%	/	/
	% GDP	0%	0,045%	0%	0%	0%
Strategy C	% TAX REVENUE	0%	0%	100%	0%	0%
	AMOUNT (10`6 €)	0	0	635	0	0
	% of current expenditure	0%	0%	1,177%	0%	0%
	% total public expenditure	0%	0%	0,092%	/	/
	% GDP	0%	0%	0,045%	0%	0%
Strategy D	& TAX REVENUE	0%	0%	0%	100%	0%
	AMOUNT (10`6 €)	0	0	0	635	0
	% of current expenditure	0%	0%	0%	0,493%	0%
	% total public expenditure	0%	0%	0%	/	/
	% GDP	0%	0%	0%	0,045%	0%
Strategy E	% TAX REVENUE	0%	0%	0%	0%	100%
	AMOUNT (10`6 €)	0	0	0	0	635
	% of current expenditure	0%	0%	0%	0%	1,170%
	% total public expenditure	0%	0%	0%	/	/
	% GDP	0%	0%	0%	0%	0,045%
Strategy F	% TAX REVENUE	50%	0%	0%	50%	0%
	AMOUNT (10`6 €)	317	0	0	317	0
	% of current expenditure	4,510%	0%	0%	0,247%	0%
	% total public expenditure	0,046%	0%	0%	/	/
	% GDP	0,022%	0%	0%	0,022%	0%
Strategy G	% TAX REVENUE	20%	20%	20%	20%	20%
	AMOUNT (10`6 €)	127	127	127	127	127
	% of current expenditure	1,804%	0,895%	0,235%	0,099%	0,234%
	% total public expenditure	0,018%	0,018%	0,018%	/	/
	% GDP	0,009%	0,009%	0,009%	0,009%	0,009%

Source: Own calculations.

ties that can mitigate the impacts of emissions. An example regarding our case of CO_2 could be reforestation.

From Table 12 we can see how spending the available state revenue in these five options seems to have substantial quantitative effects and corresponds to, alternatively:

- 9.0% of current national environmental expenditure;
- 4.5% of current state R&D investments;
- 1.2% of current expenditure in education and training;
- 0.5% of employers' National Insurance contributions;
- 1.2% of employees' National Insurance contributions.

To further analyse and compare the different state revenue spending alternatives it is possible to evaluate the effects of each investment strategy on induced demand. In order to do so, we re-run the Input-Output model including in the final demand vector both the loss of air travel demand and the increased state expenditure in the various chosen sectors. In particular, we are interested in valuing the relationship between loss and increase in induced expenditure. The next step of the analysis would be that of quantifying the aggregated effects in terms of CO_2 emission reductions to better assess the relation between the second and the first dividend. In this sense, a selective choice of the actors that should benefit by the re-investment, within the chosen revenue recycling strategy, should be able to enhance and not hinder the environmental effects of the tax. To do so there is the need to develop reliable indices of "eco-friendly" behaviour that can be used for these means.

Table 13 reports, for each of the examples of revenue spending strategies, the aggregated[21] induced demand and its *ratio* with respect to the quantity invested, not considering (first column) and considering (second column) the negative effects due to the reduced air travel demand (ΔR).

$$(7) \qquad Ratio1 = \frac{Induced\ demand\ by\ revenue\ recycling}{State\ revenue}$$

_____

[21] On all sectors.

$$(8) \ Ratio2 = \frac{Induced \ demand \ by \ revenue \ recycling + \Delta induced \ demand \ by \ tax}{State \ revenue + \Delta R}$$

Recall that the induced expenditure caused by the loss of industry revenue due to a 5% tax on domestic and continental European flights would be $-112.894 \cdot 10^6$ (equal to 0.008% of the Italian GDP).

In the case where 100% of the state revenue is invested on environment (Strategy A)[22] the sectors that remain with a negative induced expenditure are only *Air Transport* and *Other Transport*. This means that the state revenue recycling has created a larger positive effect on all the other sectors that are also negatively impacted by the loss in aviation demand (Graph 1). The sectors that gain the most by this strategy are *Environmental Services* and *Professional Activities*; the latter was one of the most hit sectors by the decrease in air travel demand.

The sectors more positively influenced by an increased state expenditure in *R&D* (Strategy B) are the *R&D* sector itself, *Chemicals and Artificial Fibres, Professional Activities, Commerce and IC&T* and the only remaining sector with a negative induced demand is that of Air transport.

An increased state expenditure in education mostly affects the sectors of *Education* and *Real Estate* (Strategy C).

Moreover, a reduction in labour taxes increases the demand for professional services and National Insurance expenditure (Strategy D and E).

The mixed strategy of spending 20% of the state revenue, due to the aviation tax, on each option (Strategy G) has a positive effect on all of the sectors in which the revenue is spent and also *Professional Activities, Chemicals and Artificial Fibres* and *Real Estate*. The total amount of induced expenditure is, in this case, lower than that produced by Strategy A, but, depending on the poli-

---

[22] In order to use the Istat I-O tables to analyse this case, we have had to introduce another simplification, that of considering investments only in environmental services, like waste and sewage management, though state investment possibilities on environment are much more diverse.

cy-maker objectives, there might be some added value due to the fact that the positive effects are more largely distributed over the different sectors of the economy.

It is important to recall that these analyses refer to the variation in expenditure due to the tax and not to the total money flows in the economy.

Purely for methodological purposes, we show how the policy-maker could start approaching the maximisation problem to choose the optimal strategy. We therefore set up an example assuming the policy-maker is interested in mixed strategies, so that only the "efficient" amount of state revenue is allocated to each option, we can for example use the following utility function:

$$
(9) \qquad u_i\left(\frac{x_i}{x_T}\right) = -w_i \frac{x_i}{x_T} \log\left(\frac{x_i}{x_T}\right)
$$

where x_i is the fraction of state revenue (x_T) invested in sector i and w_i is a parameter of the function.

In this example we use identical utility functions, with all w_i equal to one, as we are not attempting to forecast the preferences of policy-makers or society.

In these example conditions, the strategy that maximises utility is that of investing: 38.5% of the state revenue on Environmental Services; 20.5% on R&D, 0.0% on education and 41.0% reducing labour taxes. These results have no value itself, but show how the problem should be dealt with.

4. - Conclusions

This paper has analysed the complex and controversial issue of introducing an aviation carbon tax in Italy trying to develop a simple model of assessment and management of the multiple effects caused by the introduction of environmental taxes. Analysing the issue of environmental taxation in theory is more straightforward than in practice. Green taxes in the real world have to interact within a complex system of regulations and institutions (for-

TABLE 13

INDUCED DEMAND OF VARIOUS
REVENUE-RECYCLING OPTIONS

	State revenue recycling strategy	Induced demand	Total induced demand
	% TAX REVENUE	228%	229%
	AMOUNT (10^6 €)	1445	1332
Strategy A	% of current expenditure	20,542%	18,937%
	% total public expenditure	0,210%	0,194%
	% GDP	0,102%	0,094%
	% TAX REVENUE	184%	181%
	AMOUNT (10^6 €)	1169	1056
Strategy B	% of current expenditure	8,247%	7,450%
	% total public expenditure	0,170%	0,154%
	% GDP	0,082%	0,074%
	% TAX REVENUE	121%	113%
	AMOUNT (10^6 €)	771	658
Strategy C	% of current expenditure	1,429%	1,220%
	% total public expenditure	0,112%	0,096%
	% GDP	0,054%	0,046%
	% TAX REVENUE	148%	143%
	AMOUNT (10^6 €)	943	830
Strategy D	% of current expenditure	0,733%	0,645%
	% total public expenditure	/	/
	% GDP	0,066%	0,058%
	% TAX REVENUE	148%	143%
	AMOUNT (10^6 €)	943	830
Strategy E	% of current expenditure	1,737%	1,529%
	% total public expenditure	/	/
	% GDP	0,066%	0,058%
	% TAX REVENUE	188%	186%
	AMOUNT (10^6 €)	1194	1081
Strategy F	% of current expenditure	0,880%	1,580%
	% total public expenditure	/	/
	% GDP	0,084%	0,076%
	% TAX REVENUE	166%	166%
	AMOUNT (10^6 €)	1054	969
Strategy G	% of current expenditure	0,409%	0,376%
	% total public expenditure	/	/
	% GDP	0,074%	0,068%

Source: Own calculations.

mal and non formal) that influence the market and that many economic models cannot account for.

In this context, we have proposed a three-step model based on the integration of:

1. a partial equilibrium analysis to evaluate the effects of the aviation tax on the market equilibrium, on industry and state revenue, on consumer and producer welfare;

2. an Input-Output model to take into account the effects of the tax on the other sectors of the economy;

3. a Green Taxation Matrix (GTM) to analyse and compare different tax revenue management strategies.

The main drawback of using partial-equilibrium and Input-Output models is that they assume a fixed technological status of the economy and do not therefore account for changes in production strategies or preferences that instead are dynamic variables. Though for small changes, this can be considered an acceptable approximation. The advantage of this model is that it is simple, transparent and has small data requirements so it can be applicable to real cases and usable by policy-makers without the need of particularly sophisticated technical skills.

In this work, we have run a first simulation regarding a green tax on aviation in the context of Italy, using data on four different segments of the market: domestic, European, leisure flights and considering only national carriers. In particular, we have simulated the impacts of three levels of tax, corresponding to 5%, 10% and 15% of the current price and their effects when the revenue is then invested in five sectors (*environment, R&D, education and training, to reduce employer and employee National Insurance contributions*). The results of our analysis, though subject to the model and data limitations, show that, quantitatively:

– the revenue loss for the aviation sector and those related to it (by intermediate good trading) is limited;

– the state revenue generated by green taxation can be considerable and might be capable of producing important effects on the market;

And qualitatively:

– there is an interesting increase in the variety of possible uses of the revenue that can target different objectives.

For example a 5% tax on all flights to/from Italy to/from all domestic and continental European (EU+ non EU States) destinations produces a 2.3% increase in the unit price and a 2.7% decrease in the quantity demanded. The revenue of the aviation sector is reduced by 0.4% ($52 \cdot 10^6$ €) which generates a loss of demand on the whole economy of $113 \cdot 10^6$ €. The State collects revenue for $635 \cdot 10^6$ €, that corresponds respectively to (alternatively):

– 9.0% of current national environmental expenditure;
– 4.5% of current state R&D investments;
– 1.2% of current expenditure in education and training;
– 0.5% of employers' National Insurance contributions;
– 1.2% of employees' National Insurance contributions.

Because of the interdependencies present in the economy, the money invested by the state in these options is also capable of producing even bigger effects by means of induced demand in other sectors: respectively 228% (100% investment in environment), 184% (R&D), 121% (education) and 148% (National Insurance contributions) of the state tax revenue.

Indeed, when analysing the combined effects of a reduced demand of air travel and an increased state expenditure in the five sectors of the example, for what concerns money flows, we find that the benefits from government expenditure in all cases outweigh, in aggregate terms (and also in most single sectors except for *Air Travel*), the loss of induced expenditure due to the tax on the aviation sector.

Our simulations show that an aviation tax in Italy could generate a revenue that would allow quite substantial reductions in labour taxes (a 5% carbon tax allows a 0.5% cut of employers' National Insurance contributions), though literature suggests that this does not guarantee the existence in practice of an "employment second dividend". In particular the Italian labour market is characterised by a strong rigidity that might cut down these positive potential effects.

Our model is built to also analyse other tax revenue recycling options, including spending on environment related activities, though literature is not in total agreement on earmarking policies. Our objective is, nevertheless, to propose an instrument (the GTM) that is capable of comparing all combinations (over and above those examined in our case-study) in order to help policy-makers choose the most efficient allocation of the tax revenue; though the first draft of our Green Taxation Matrix does not take into account the administrative costs of tax management.

From the results of our simulations, we find that in Italy the effects of spending the revenue on Environment and R&D can be quantitatively considerable (a 5% tax collects respectively 9.0% or 4.5% of current expenditure). This revenue recycling option – out of the classic double dividend framework – could therefore be interesting as Italy suffers from low investments in these strategic areas. If current state investments in these sectors are inefficient, the allocation of the revenue to these sectors could also generate long-term positive effects on innovation, competitiveness and even the development and diffusion of environmental culture.

A further step of the analysis would be that of increasing the sophistication of the GT matrix to include a quantitative evaluation also of the effects of the different strategies on CO_2 emissions to be able to compare the "second dividends" also in the light of their influence with respect to the effectiveness of the first dividend, the environmental one. In this context, a selective choice of the activities that should benefit from the re-investment – aimed at favouring "low-emitters" – might enhance the environmental effects.

In addition to that said above, there is another interesting "extra" effect of green taxes, that is public acceptance. Green taxes, in the light of the double dividend or of these "multiple" effects, present characteristics that can potentially obtain a certain amount of public favour; this should be recognised and enhanced by policy-makers, through new tax policies based on higher transparency. The demand for environmental protection and responsibility, though subject to high elasticity, is in sustained growth among the public. Acceptance of green taxes would probably go together with tax shifting policies. These policies, that are intended to shift the burden of taxation from "goods" to "bads", have the need of structural changes that necessarily generate costs.

To summarise, even if the classical fiscally-neutral double dividend might be moderate, the tax multiple dividends and potential long-term effects can make environmental taxes a reasonable policy instrument, if efficiently managed.

BIBLIOGRAPHY

ANDERSEN M.S., *Governance by Green Taxes: Making Pollution Prevention Pay*, Issues in Environmental Politics, Manchester University Press, Manchester, UK, 1994.

ASSOCIATION OF EUROPEAN AIRLINES, *Yearbook 06*, Association of European Airlines, 2006.

BOSELLO F. - CARRARO C., «Recycling Energy Taxes: Impacts on a Disaggregated Labour Market», *Energy Economics*, Vol. 23, no. 5, 2001, pages 569-594.

BOSQUET B., «Environmental Tax Reform: Does It Work? A Survey of the Empirical Evidence», *Ecological Economics*, Vol. 34, 2000, pages 19-32.

BOVENBERG A.L. - DE MOOIJ R.A., «Environmental Levies and Distortionary Taxation», *The American Economic Review*, Vol. 84, 1994, pages 1085-1089.

BRONS M. - PELS E. - NIJKAMP P. - RIETVELD P., «Price Elasticities of Demand for Passenger Air Travel: A Meta-Analysis», Holland, Tinbergen Institute, *Discussion Paper*, no. 3(047), 2001.

BROUWER R. - BRANDER L. - VAN BEUKERING P., «A Convenient Truth: Air Travel Passengers' Willingness to Pay to Offset Their CO_2 Emissions», *Climatic Change*, Vol. 90, no. 3, 2008, pages 299-313.

CARRARO C. - GALEOTTI M., *Ambiente, occupazione e progresso tecnico: un modello per l'Europa*, Bologna, Il Mulino, 1995.

CARRARO C. - GALEOTTI M. - GALLO M., «Environmental Taxation and Unemployment: Some Evidence on the Double Dividend Hypothesis», *Journal of Public Economics*, no. 62(1/2), 1996, pages 141-181.

CARRARO C. - SINISCALCO D., *Environmental Fiscal Reform and Unemployment*, Kluwer Academic Publishers, Dordrecht, DK, 1996.

CONRAD K. - A. LOSCHEL A., «Recycling of Eco-taxes, Labor Market Effects and the True Cost of Labor - A CGE Analysis», *Journal of Applied Economics*, Vol. VIII (2), 2005, pages 259-278.

DINGS J.M.W. - WIT R.C.N. - LEURS B.A. - DAVIDSON M.D. - FRANSEN W., *External Costs of Aviation*, Delft, Centre for Energy Conservation and Environmental Technology, Research Report, no. 299, 2003, pages 96-106.

ECON, *GHG Emissions from International Shipping and Aviation*, ECON Report, no. 38400, 2003.

— - —, *The Political Economy of the Norwegian Aviation Fuel Tax*, report for Joint Meeting of Tax and Environment Experts, OECD, 2005.

ENAC, *Annuario statistico 2005*, Ente Nazionale per l'Aviazione Civile, 2006.

EUROPEAN COMMISSION, «Tax-based EU Own Resources: An Assessment», Brussels, European Commission, *Working Paper*, no. 1/2004, 2004.

EUROPEAN ENVIRONMENTAL AGENCY, *Environmental Signals 2001*, European Environmental Agency Environmental assessment report, no. 8, 2001.

EUROPEAN PARLIAMENT, *European Parliament Resolution on Reducing the Climate Change Impact of Aviation*, Strasbourg, European Parliament, Procedure number INI (2005) 2249, OJ C 303E, 2006, pages 119-123.

GOULDER L.H., «Environmental Taxation and the Double Dividend: A Reader's Guide», *International Tax and Public Finance*, Vol. 2, no. 2, 1995, pages 157-183.

INFRAS - ECOPLAN, *Economic Impact Analysis of Ecotax Proposals - Comparative Analysis of Modelling Results*, Final Report, 1996.

INTERNATIONAL MONETARY FUND, *Government Finance Statistics Yearbook 2007*, vol. 31, International Monetary Fund, 2007.

IPCC - PENNER J.E. - LISTER D.H. - GRIGGS D.J. - DOKKEN D.J. - MCFARLAND M., *Aviation and the Global Atmosphere*, UK, Cambridge University Press, 1999, page 373.

ISTAT, *I consumi delle famiglie*, Annuari ISTAT, no. 12, 2007a.

— - —, *Conti economici nazionali*, ISTAT, 2007b.

KEEN M. - STRAND J., «Indirect Taxes on International Aviation», *Fiscal Studies*, Vol. 28, no. 1, 2007, pages 1-41.

LANZA A. - SANMARCO G., «Strumenti per la politica ambientale: il caso della carbon tax in italia», in MUSU I. (eds.), *Economia e ambiente*, Bologna, Il Mulino, 1993, pages 222-223.

MARKANDYA A., «Green Taxation», in SAXE H. - RASMUSSEN C. - REINHARD A.J. (eds.), *Green Roads to Growth: Proceedings of Expert and Policy Maker Forums*, held in Copenhagen 1-2 March, 2006, Copenhagen, Environmental Assessment Institute, 2006.

OECD, *Economic Instrument Database 2007*, OECD, 2007.

PEARCE B. - PEARCE D., «Setting Environmental Taxes for Aircraft: A Case Study of the UK», *CSERGE, Working Paper*, CSERGE 2000-26, 2000.

PEARCE D.W., «The Role of Carbon Taxes in Adjusting to Global Warming», *The Economic Journal*, Vol. 101, no. 407, 1991, pages 938-948.

PESZKO G., «Integrating Public Environmental Expenditure Management and Public Finance in Transition Economies», in *Proceedings of Fifth Expert Group Meeting on Finance for Sustainable Development: Testing New Policy Approaches*, Nairobi, Kenya, 1-4 December 1999, 1999.

PEZZEY J. - PARK A., «Reflections on the Double Dividend Debate», *Environmental and Resource Economics*, Vol. 11, no. 3, 1998, pages 539-555.

R DEVELOPMENT CORE TEAM, *R: A Language and Environment for Statistical Computing*, Vienna, Austria, R Foundation for Statistical Computing, 2007.

REPETTO R. - SCHWARTZ J., «Nonseparable Utility and the Double Dividend Debate: Reconsidering the Tax-Interaction Effect», *Environmental and Resource Economics*, Vol. 15, no. 2, 2000, pages 149-157.

ROBERTSON J., «Eco-Taxes», *New Internationalist*, Vol. 278, 1996, pages 28-30.

SEWILL B., *The Hidden Cost of Flying*, Aviation Environment Federation, London, 2003.

SMITH S., *"Green" Taxes and Charges: Policy and Practice in Britain and Germany*, London, The Institute for Fiscal Studies, 1995.

VARIAN H.R., *Intermediate Microeconomics - A modern Approach*, 2nd ed., New York and London, W.W. Norton & Company, 1990.

WEIZSACKER E.U. - JESINGHAUS J., *Ecological Tax Reform - A Policy Proposal for Sustainable Development*, London & New Jersey, ZED Books, 1992.

WIT R.C.N. - DINGS J.M.W. - MENDES DE LEON P. - THWAITES L. - PEETERS P. - GREENWOOD D. - DOGANIS R., *Economic Incentives to Mitigate Greenhouse Gas Emissions from Air Transport in Europe*, Delft, Centre for Energy Conservation and Environmental Technology, 2002.

Is it Time for a Revival of ETR in Italy? Energy Elasticities and Factor Substitutability for Manufacturing Firms

Rossella Bardazzi - Filippo Oropallo - M. Grazia Pazienza*

University of Florence   Istat, Rome   University of Florence

Environmental Tax Reforms (ETRs) have recently enjoyed renewed widespread attention as a tool aimed at GHG emission abatement. In this paper the efficacy of energy-related taxes in relation to Italian firms' behaviour is analyzed. A translog model is used to estimate factor demands for a panel of industrial firms divided between small-medium and large enterprises. Our analysis estimates large and negative own price elasticities for energy. As for other inputs, we find complementarity between energy and capital and substitutability between energy and labour for small firms therefore a double dividend effect could take place if ETR is implemented. [JEL Classification: C33, D24, H23, Q41].

Keywords: environmental tax reform; energy factor demand; micropanel data; translog.

1. - Introduction

Environmental Tax Reforms (ETRs) have recently enjoyed renewed widespread attention for two main reasons. On the one hand, the recent difficulties in reaching a worldwide agreement on policies to combat climate change call for a shifting of responsibility to individual countries. Among the many instruments available for combating GHG emissions, environmental taxes and

---

* <rossella.bardazzi@unifi.it>; <oropallo@istat.it>; <mariagrazia.pazienza@unifi.it>.

other economic instruments have demonstrated their efficacy, notwithstanding the fact that their distributional impact needs to be seriously evaluated.

On the other, at the present time the general attitude of governments is to shift taxes away from personal income and capital towards taxes on goods and consequently mainly on consumption, energy use as well as pollution[1]. This choice has a dual result (a win-win hypothesis or a double dividend) of reducing pollution and the disincentive effect caused by income tax on labour or savings and, at the same time, creating a certain degree of fiscal illusion[2] — which, in the case of environmental taxes, may be relevant — given the rising political acceptability of such taxes due to general concern regarding climate change[3].

A first wave of ETRs, originating in Nordic countries in the first half of the Nineties, has been analyzed in depth in economic literature[4]. These reforms came as a response to the European Union's inability to establish a general policy instrument — a European Carbon Tax — although they were strongly encouraged to do so by the Commission itself after Delor's White Paper. Funda-

---

[1] Goods are not all the same: ETR must be designed to shift the tax base away from taxing "good" resources such as investment and labour, towards taxing "bad features" such as pollution and inefficient resource use. Essential characteristic of an ETR is therefore revenue recycling aimed at achieving a double dividend. However, economic literature has stressed that the positive effect known as double dividend may be outweighed by the "tax interaction effect", the welfare cost arising from the increases in output prices due to energy taxes.

[2] Fiscal illusion can be defined as the idea that taxpayer considers his tax burden as smaller than it really is. Environmental taxes are usually unit taxes and for this reason can be perceived as incorporated into the product price.

[3] Generally speaking command and control regulation has been considered easier to employ by policy makers – in a public choice framework - because of the lesser transparency of costs and rent-seeking behaviour of polluting firms. The main argument is that with an environmental tax there is a transfer from polluters to the government, while, under a regulation regime, pollution rights on the remaining emissions stay with firms. Therefore, if the polluters cannot expect the tax revenue to be recycled with subsidies, they prefer command-and-control instruments. However, following increasing concern about environmental quality and sustainability growth, green taxes have been more favourably considered. On this issue see PEARSON M. (1995) and OECD (2006).

[4] The first countries to implement these ETRs were Finland, Sweden, Denmark in the early 1990s, the Netherlands and UK in the mid-1990s and Germany in 1999. See, among others, EKINS P. - SPECK S. (1999, 2007); BARDAZZI R. - OROPALLO F. - PAZIENZA M.G. (2004).

mentally, all countries restructured the existing taxes on energy products or introduced new ones but with a different magnitude of shifting among tax bases and different ways of recycling revenue.

A second wave of ETRs has been discussed and implemented in Europe since the early 2000s. Some Eastern Countries, such as Estonia and Czech Republic introduced an ETR based on the Scandinavian model; Germany and the UK made significant adjustments in tax rates. In this period a new Energy Taxation Directive set minimum tax levels for energy products in 2003, but tax rates still differ noticeably among member countries. However, not all fiscal changes labelled as ETRs have been properly designed in the past. Some of them have proved too ambitious (a carbon levy – under an ETR framework – has been under discussion in Ireland and France since the late Nineties) and have shown a lack of time consistency[5] while others have been based more on revenue concerns than on environmental efficacy.

As shown by a draft proposal and recent public meetings[6], the European Commission tried once again in 2009 to put carbon taxation at the top of agenda as an additional instrument for fighting climate change. The draft proposal analyzed different policy approaches which ranged from simply avoiding the negative effect of overlapping instruments (such as energy taxes and EU-Emission Trading Scheme (ETS)) to the introduction of "an additional uniform CO_2-related tax at the top of taxes already existing under European Energy Directive to complement EU-ETS" (p. 5). As a new common tax appears politically unfeasible, the Commission is evaluating how to amend the common energy taxation policy (the base level tax rate set in Energy Taxation Directive) in

---

[5] In some cases the policy targets were too ambitious and governments were forced to change the announced path of reform. A substantial failure of the process was experienced in Ireland, France and, as discussed later, in Italy in 1999-2000. The breakdown of the reforms emerged after the sharp rise of fuel prices leading to a sort of "fuel revolt". This was probably also connected to a wrong policy communication strategy and conflicts between the administrations involved in the projects. However, with the aim to reduce public budget crisis, Ireland introduced a carbon-based energy taxation in 2010.

[6] European Commission (2009).

order to tax energy products with reference to climate change emissions, in addition to their energy content. At the same time the revision would exempt all sectors involved in the Emission Trading Scheme and abolish tax concessions for some sectors[7].

Whatever the evaluation of the efficacy and the feasibility of a global ETR, an environmental policy aimed at GHG emission abatement is necessary for all countries, and taxing "bad features" can lead to environmental improvement. Moreover, in the case of Italy, the aim is twofold: on the one hand, all the recent projections have highlighted that Italy will be unable to meet the Kyoto target as identified by European Burden Sharing Agreement without substantial policy correction; on the other, a key goal is to cut energy dependency on imports by reducing energy intensity or changing the energy input mix.

In this paper we analyze the efficacy of energy-related taxes in relation to firms' behaviour in Italy by estimating demand elasticities and input substitution elasticities in a translog model for a panel of industrial firms. The paper is organized as follows: the first section contains a brief review of energy tax efficacy followed by some further background regarding the Italian situation (Section 2). The factor demand model used in the empirical analysis is presented in Section 3. A description of the dataset and the characteristics of energy demand by Italian firms follows in Section 4. A discussion of the results concludes.

2. - The Efficacy of Energy Taxes

A general evaluation of environmental taxes as a policy instrument can be made on the basis of various criteria. As an example of market based instruments, environmental taxes are generally characterized by economic efficiency (both in static and dynamic terms), environmental effectiveness and relatively low monitoring and administrative costs. On the downside, the effect on firms' competitiveness and possible regressive impact on house-

---

[7] For a comment linked to the proposal see LAURENT E. - LE CACHEX J. (2009).

hold income distribution have been a source of major concern for policy-makers when considering potential applications of taxes to energy and other goods[8].

As with any other policy evaluation, the environmental effectiveness of taxes can only be assessed after clarification of the expected objectives but the choice of tax base, tax rate[9] and the availability of substitutes permitting a change in the agent's behaviour are key factors to the environmental success of a policy[10].

To assess the efficacy of energy-related taxes on firms, economic literature has focused on price demand elasticities and input substitution. It is worth emphasising that an energy-related tax may induce a decline in output, an input substitution (substitution of energy with other inputs or substitution of dirty with cleaner inputs) or specific investments aimed at emission abatement. If taxation is uniform among energy inputs, firms have a general incentive for energy-saving but there is no direct incentive to reduce emissions by inter-fuel substitution as firms cannot reduce the taxes they have to pay by changing their input mix or by making specific, low-emission investments. If the tax is designed according to differences in input characteristics (such as energy or carbon content) substitution among inputs will also be activated[11].

---

[8] However, as pointed out by AGNOLUCCI P. (2004), it is very difficult to distinguish between the effect of taxes and their effectiveness: assessing the efficacy of a tax implies "ascertaining the effects of tax in relation to the expected objectives and targets or to other instruments".

[9] If the aim is to resolve a pure externality problem, as in the pigouvian framework, the crucial point is the quantification of the externality in order to set the proper tax rate.

[10] Another crucial element to take into account is the existence and the range of exemptions.

[11] It is worth recalling that energy taxes are based on the quantity of energy consumed, whereas the tax design of a carbon tax should be based only on emissions or on the CO_2 content of fossil fuels. In other words an energy tax is imposed on fuels (both fossil fuels and carbon-free sources such as nuclear power and renewables), while a carbon tax is restricted to carbon-based fuels only. This difference explains why a carbon tax is believed to be more efficient in reducing emissions: an energy tax works through the general cost-saving mechanism, while a carbon tax adds the fuel substitution channel to the cost-saving channel. A carbon tax equalizes the marginal abatement cost across fuels, and in this way minimises the total cost of abatement.

The direct price elasticity of energy inputs measures the re-activity to price changes and indirectly measures the potential for environmental taxes. Empirical economic literature has a long history of energy elasticity estimation and several techniques and data types have been used, even though industry energy demand estimations are less common. As a general finding, energy demand is sensitive to price change especially in the long run. As an example of a recent cross-countries study for industrial use, Liu (2004) estimated energy product demand elasticities for a panel of OECD countries and found that electricity and gas are the most significant energy products and the most sensitive to price changes[12].

A second branch of literature has focused on the complementarity or substitution relations between energy and other inputs. This is a key issue in evaluating the effect of energy-related taxes: if capital and energy were substitutes an increase in energy prices would lead to a rise in capital demand, whereas if complementarity is found, an incentive for an energy saving technology would jeopardize investment and capital accumulation.

This topic has been central to empirical literature since the oil shock of 1973 but, in spite of the abundance of empirical estimations, results have been controversial. A recent meta analysis provided by Koetse, de Groot and Florax (2008) shows that the source of variability must be attributed to the data, estimation technique, time period and geographic area. However, they found that the opportunity for capital-energy substitution is large for all regions (though estimates are substantially larger in North America than in Europe) and time periods, especially in the long run[13]. Generally speaking, many studies highlighted that estimations based on time-series found complementarity between energy and capital because they identified short-run effects, while cross-na-

---

[12] Liu's estimation confirmed higher elasticities in the long run, although the level of estimated values is lower than in previous studies. He found a value of –0.224 for gas and –0.04 for electricity in the long run.

[13] As regards Italy, MEDINA J. *et al.* (2001) working on aggregate data, found the existence of a consistent substitution between energy and labour, but weak energy-capital substitutability.

tional estimations found substitution because they measured long run potential. As highlighted by Solow (1987), micro data must be preferred because factor substitution is essentially a micro-economic phenomenon and is best analyzed with micro data. However, few estimations have been conducted on microdata and, to our knowledge, there are no previous studies using microdata to investigate this issue for Italian firms. Nguyen and Streitwieser (2008) found that energy and capital were substitutes by estimating several elasticities on a cross section of about 10,000 USA firms for 1991[14]. Arnberg and Bjorner (2007), on the contrary, used a panel of Danish industrial firms (from 1993 to 1997) and found complementarity between energy and capital. Considering the relatively low level of their estimated elasticities, they cast some doubts on the efficacy of environmental taxation.

3. - The Italian Case

Although Italy has not been a particularly active ETR country, an attempt to redesign the tax system according to the environmental goal was made in the late Nineties, known as the Carbon tax reform.

The Carbon tax reform, which came into force in 1998 (law 448/98), is a clear example of political and institutional difficulties, as the policy started with a general and coherent design and was halted after two years without any policy substitution. The aim of this reform was to reduce environmentally damaging inputs by re-shaping the previous energy-related tax rates and including coal and other high-emission energy products. Moreover, the Italian carbon tax reform was designed as a fiscally neutral reform: the increase in revenue was offset by a decrease in existing social contribution rates. In other words the reform was based on the "double dividend" hypothesis, where the first dividend was the supposed fall in emissions and the second dividend an increase in full time

---

[14] NGUYEN S.V. - STREITWIESER M.L. (2008) also highlighted a sensitivity of estimates to different definitions of elasticity.

employment through a cut in the labour tax wedge[15]. Indeed, the fiscal neutrality approach helped to alleviate concerns regarding competitiveness and to increase the political acceptability of the reform. The tax rates originally foreseen in 1999 were supposed to gradually increase up to a target level in 2005[16]. Unfortunately, in a framework of rising international energy prices, the original tax outline was never implemented and in fact tax rate revisions have been frozen at 2001 level. An *ex ante* estimation of the impact of the full reform on manufacturing firm's competitiveness was performed with a microsimulation model by Bardazzi, Oropallo, Pazienza (2004)[17] and showed an overall negligible effect on profitability (–0,6% for Gross Operating Surplus). However, these effects appeared highly differentiated according to sector and firm size due to the variability of energy expenditure as a component of intermediate costs and of the share of labour costs[18].

After this policy attempt, some minor variation of tax rate levels was implemented and as a consequence the effective tax rates of most energy products declined sharply as net prices increased.

Graph 1 highlights the opposing trends in Italy and Europe regarding energy intensity and energy taxation. The European Union shows a marked decrease in energy intensity and an overall increase in the energy implicit tax rate[19]. On the contrary, Italy shows a remarkable contraction of the energy implicit tax rate and very little improvement in energy intensity: although Italy has, historically, been characterized by very low energy intensity of GDP – one of the lowest among western developed countries – it

---

[15] Moreover, a form of earmarking, through subsidies to environmentally-related investments, was designed.

[16] The structure of excise rates has been designed in such a way that each tax rate is the sum of two components. The first term satisfies the European minimum levels, while the second is linked to the environmental objective and reflects the carbon content of each energy product.

[17] BARDAZZI R. - OROPALLO F. - PAZIENZA M.G. (2004).

[18] The *ex ante* evaluation referred to the comparison of the effective tax rates in 2000 with the hypothetical tax rates in 2005 if the original tax designed had been implemented.

[19] Despite the important energy and industrial transition process in Eastern European Countries, it is important to stress that the same trend also characterized the EU-15 aggregate.

shows one of the smallest contractions of the period (–3% as opposed to –23% on average in OECD countries), probably due to
the higher marginal abatement cost and the fact that tax rates declined.

GRAPH 1

ENERGY IMPLICIT TAX RATES* AND ENERGY INTENSITY**
IN ITALY AND EU-25 (1996 = 100)

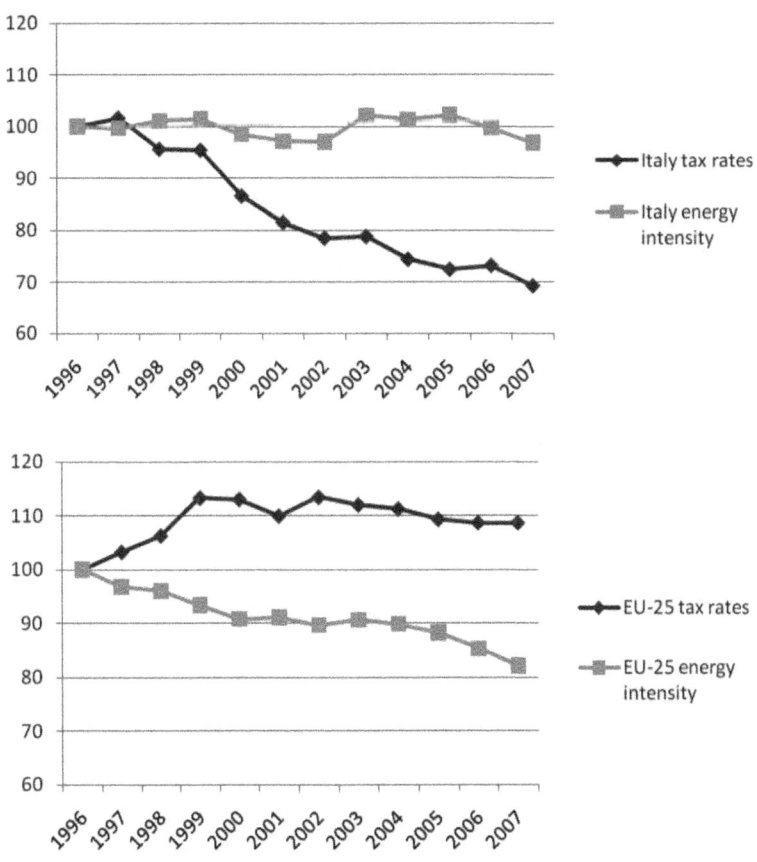

* *Ratio* of energy tax revenues to final energy consumption, deflated.
** Gross inland consumption of energy divided by GDP (kilogram of oil equivalent per 1,000
Euro).

Source: Authors' calculation on Eurostat database.

The overall reduction in the energy implicit tax rate — computed in relation to GDP — stems from a general tendency to an almost constant unit tax rate and increasing product prices. As shown by Graph 2, implicit tax rates on gas and electricity at an industrial level have declined sharply in the last decade: despite some variability linked to price variation, between 1999 and 2007 the implicit tax rates show a fall of about 40%.

GRAPH 2

IMPLICIT TAX RATES FOR OVERALL ENERGY* AND
ELECTRICITY** AND GAS*** FOR INDUSTRIAL USE (1999 = 100)

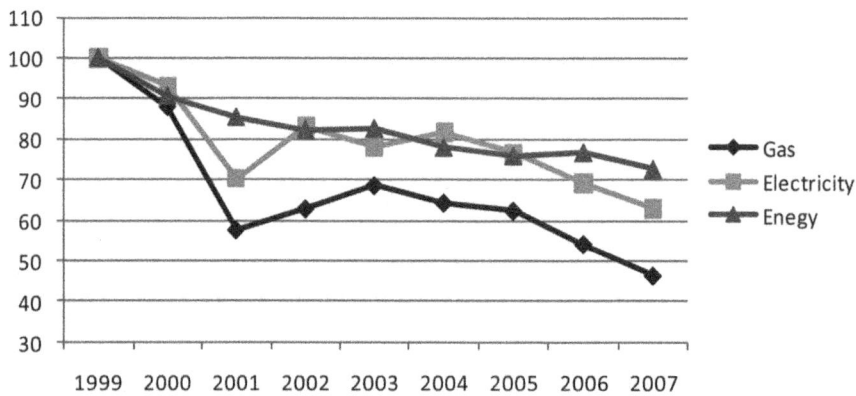

* *Ratio* of energy tax revenues to final energy consumption, deflated.
** *Ratio* of tax rate (state and local) to product price for industrial use below 200,000 kWh/month.
***Ratio of tax rate (state and local) to product price for industrial use below 1,200 thousand of Mc.

Source: Eurostat for prices and D.Lgs. 504/1995 for tax rates.

The lack of inflation adjustment of tax rates has transformed Italy from a high energy tax country into a mid-range country, at least as regards business use taxation. According to European Commission comparative data, Italy has a very high tax rate for gasoil, mid range for petrol and liquid petroleum gas, and quite low for other relevant energy inputs such as electricity, gas, coal and others[20]. Moreover — as for other European partners — there

---

[20] EUROPEAN COMMISSION (2010).

are a number of tax concessions for specific sectors such as agriculture and transport.

This trend showed that energy tax instruments have not been managed actively and properly in recent years and unfortunately a general environmental policy aimed at reducing climate damaging emissions or increasing energy saving innovations is still lacking.

4. - The Factor Demand Model

In this paper we use the translog model developed by Christensen *et* al. (1973), widespread in energy elasticity estimations. This is due to the fact that the translog function belongs to the class of "flexible functional forms" which can approximate an arbitrary function up to the second order, although some properties are verified only in a finite set of parameters.

The translog model assumes a general indirect cost function (given the equivalence of production and cost functions) and applies Shephard's lemma to determine the demand functions of the factors and the share equations. The *n*-equation system of input factor shares to be estimated can be written as:

$$(1) \qquad S_i = c_i + \sum_{j=1}^{n} \gamma_{ij} \ln p_j \qquad j = 1,\ldots i,\ldots n$$

where "*j*" are the *n* factor inputs, S_i is share of factor *i* on total cost, and *p* are input prices.

To avoid a singular covariance matrix, one equation must be dropped from the system. By dividing all the prices in the remaining set of equations by the price of the *k* dropped input, it is possible to omit that equation. So, the dropped input (for instance capital) is used as the *numeraire* and the parameters of its equation are calculated using the summing, price homogeneity and symmetry conditions as constraints.

$$(1\text{-}bis) \qquad S_i = c_i + \sum_{j=1}^{n-1} \gamma_{ij} \ln p_j / p_k \qquad j = 1, \dots i, \dots n\text{-}1$$

As usual, this kind of analysis focuses on elasticity estimation in order to assess the magnitude of reaction to the policy signal. Price elasticities can be obtained by deriving each cost share with respect to the own price[21]:

$$(2) \qquad \frac{\partial S_i}{\partial p_i} = \frac{\gamma_{ii}}{p_i} - \beta_i \frac{\partial \log p}{\partial p_i}$$

Simplifying, we may arrive at the elasticity formula of the input in relation to its own price:

$$(3) \qquad \eta_{ii} = \frac{\gamma_{ii} + S_i^2 - S_i}{S_i}$$

While, for the cross-price elasticity of input i with respect to p_j we have[22]:

$$(4) \qquad \eta_{ij} = \frac{\gamma_{ij} + S_i S_j}{S_i}$$

Our model considers four input share equations (labour, material, energy and capital) with a pooling data (longitudinal and cross sectional) estimation and to avoid singularity we have decided to drop the capital demand equation. Each equation is a function of the p_{jt} price of inputs (j = *lab, mat, ener, cap*) and of the level of real output Y_t/P_t. Moreover, in order to consider the non-homotheticity of the underlying production function (*i.e.* not constant returns to scale), the equation system has been integrat-

---

[21] See THOMPSON H. (2006) for a formal derivation of the formulas.

[22] As shown by THOMPSON H. (2006) there are several ways of computing factor substitution. Divergence of results between studies may depend on the measure used. However, Allen elasticities are considered inappropriate in the case of more than two production factors, while Morishima and cross-price elasticities are considered reliable by the majority of authors.

ed with the following instrumental variables: (1) the logarithm of the number of workers as a proxy of the size of the firm (emp_{it}), (2) the year dummies (to capture calendar effects and technical progress) and (3) sectoral dummies (DS_{ij}) (to capture the individual effects of each industry). The modified version of equation *(1)* is the following:

$$S_{it} = d_i + \sum_j \gamma_{ij} \ln p_{jt} / p_{kt} + \beta_i \ln\left(Y_t / P_t\right) +$$

(5)
$$+\delta_i \ln\left(emp_t\right) + \sum_t \tau_{it} year_{it} + \sum_{j \in Nace} \vartheta_{ij} DS_{ij} + u_{it}$$

$$\begin{cases} i = lab,\ mat,\ ener,\ cap \\ t = panel\ observ. \end{cases}$$

Parameters have been estimated using the SUR technique[23] which allows correlated errors between equations and uses GLS to estimate parameters in a more efficient way. Then we imposed the usual restrictions on coefficients:

Adding up $\sum_i d_i = 1, \sum_i \beta_i = 0$ and $\sum_i \gamma_{ij} = 0$;

(6) Homogeneity $\sum_j \gamma_{ij} = 0$;

Symmetry of substitution effects $\gamma_{ij} = \gamma_{ji}$.

These restrictions, as shown by Thompson (2006), may improve estimation robustness.

As regards the constant term, with panel data we can take into account the unobserved firm heterogeneity in input choice behaviour by allowing the intercepts to be different for each firm.

---

[23] ZELLNER A. (1962).

5. - Data Description

In this paper we utilize a micro-dataset and a micro-simulation model for Italian firms built within a European project called DIECOFIS[24]. The dataset used by this microsimulation model is called EISIS (*Enterprise Integrated and Systematized Information System*) which is a multi-source business data bank based upon microdata created at the Italian National Statistical Office[25]. In particular, to model energy taxes and fuel consumption by firm, data from the *Manufacturing Product Survey* (*Prodcom*) was matched with the main database[26]. The resulting data covers all Italian manufacturing firms with more than 19 workers and a sample of small firms with more than 2 and less than 20 workers. These data are available for the years 2000-2005 and include information about expenditures (net of VAT) and consumption in physical units of several energy sources[27]. In the model, specific tax rates *per* economic activity are considered and both energy prices and taxes are determined at the firm level. To estimate our demand system for manufacturing firms information on labour, capital and raw material costs and prices was necessary. While

---

[24] DIECOFIS (Development of a System of Indicators on Competitiveness and Fiscal Impact on Enterprise Performance) is a project financed by the Information Society Technologies Programme (IST-2000-31125) of the European Commission and coordinated by the Italian National Institute of Statistics. This model is designed to analyse the effect of fiscal policy on enterprises, considering social contributions, corporate taxes, and energy taxes. DIECOFIS has been used to perform several policy evaluations to monitor (*ex ante/ex post*) the effectiveness of policies. The model is run at ISTAT where data is produced but the Institute bears no responsibility for analysis or interpretation of the data. See BARDAZZI R. - PARISI V. - PAZIENZA M.G. (2004) for an overview of the main features of the model.

[25] The integrated and systematized information system on enterprises is the result of an integration process of different administrative sources. The statistical register of Italian active enterprises (ASIA) has been used as a "spine" for this integration process. On this information, other sources have been attached: Large Enterprise Accounts (SCI); Small and Medium Enterprise Survey with less than 100 workers (PMI); Foreign Trade Archive (COE); other surveys such as the Community Innovation Survey (CIS) and the ICT Survey. All of the above ISTAT surveys are based on common EUROSTAT standards and classifications.

[26] A statistical matching procedure was implemented as described in BARDAZZI R. - OROPALLO F. - PAZIENZA M.G. (2004).

[27] Electricity, coal, LPG, diesel, gasoline, metallurgic coke, petroleum coke, fuel oil, natural gas, and other minor products.

factor costs at the firm level are available from EISIS, the price of labour was computed from the dataset as the firm's total personnel expenses *per* hour worked and the price of energy was computed from the model as the firm's weighted sum of the prices, after taxes, of the most important eight energy sources. Finally, the prices of materials and capital were computed on a national account basis respectively as the price index of materials from supply and use tables (SUTs) at a two digit level of the NACE classification[28], and the price index of gross and net capital for each branch of economic activity[29]. Finally, the real output of equation *(5)* is computed as the firm production value from the dataset deflated by the sectoral price index of output from SUTs.

For this study a balanced panel of firms was created using the EISIS dataset. From the available data, we were left with 21,450 observations covering 3,575 manufacturing firms which were surveyed for all six years, of which roughly 85 per cent had at least 100 workers. Therefore, our panel was biased towards large firms since the survey covered the universe of these enterprises. However, this feature of our data does not diminish the significance of our findings as the firms in our panel represent on average about 35 per cent of total industrial gross value added (GVA), 44 per cent of employment and 40 per cent of energy expenditure. Firms are distributed across manufacturing sectors as shown in the first column of Table 1 with a third of our sample allocated in the very energy-intensive metallurgic industry (divisions 27, 28 and 29) which closely follows the distribution of the universe of Italian manufacturing firms.

The majority of sectors showed fairly stable energy consumption over the period: besides the Mining activities — which are not very developed in Italy — other energy intensive activities are Textiles, Food Products, Pulp and paper products, and the Chemical and Plastic industries. For these sectors energy costs represent a share ranging from 5 to 10 per cent of their production costs. Finally, the median energy expenditure is very differ-

---

[28] ISTAT (2010), *Il sistema delle tavole input-output.*
[29] ISTAT (2009), *Conti economici nazionali.*

TABLE 1

SOME CHARACTERISTICS OF THE FIRM PANEL
(AVERAGE 2000-2005)

Economic activities	(%)	Energy intensity[a]	Energy expenditure over total costs	Median energy expenditure (total = 100)[b]	
				2000	2005
13, 14 - Metal ores and other mining	0.79	0.471	0.186	394	545
15, 16 - Food products, beverages and tobacco	9.00	0.331	0.027	206	250
17 - Textiles	7.20	0.389	0.053	112	120
18 - Wearing apparel	3.20	0.118	0.023	24	15
19 - Luggage, handbags, and footwear	2.28	0.108	0.017	65	30
20 - Wood, except furniture	2.01	0.302	0.036	88	90
21 - Pulp, paper and paper products	2.76	0.387	0.048	224	240
22 - Publishing, printing and reproduction of rec. media	2.66	0.083	0.018	24	20
24 - Chemicals and chemical products	7.28	0.268	0.035	135	165
25- Rubber and plastic products	5.73	0.283	0.033	471	460
26 - Other non-metallic mineral products	6.66	0.690	0.101	253	315
27 - Basic metals	5.25	0.546	0.052	365	560
28 - Fabricated metal products	9.63	0.225	0.029	124	160
29 - Machinery and equipment n.e.c.	15.61	0.061	0.012	112	80
30 - Office machinery and computers	0.23	0.047	0.015	6	5
31 - Electrical machinery and apparatus n.e.c.	4.30	0.095	0.014	47	45
32 - Radio, television and communication equipment	1.89	0.077	0.013	18	15
33 - Medical, precision and optical instruments	2.65	0.059	0.013	41	40
34 - Motor vehicles, trailers and semi-trailers	4.05	0.143	0.020	353	155
35 - Other transport equipment	1.76	0.086	0.015	35	25
36 - Furniture; manufacturing n.e.c.	5.05	0.094	0.014	59	70
Total	100	0.247	0.033	100	100

[a] Energy consumption in Toe/value added in thousand euros. The energy products considered are the 8 most used by industrial firms: electricity, natural gas, gasoil, fuel oil, LPG, gasoline, coal, and coke.
[b] Sectoral energy median with respect to the panel median.

ent across sectors and it is a useful measure to take into account the technological characteristics and the dimensional structure of each production. As expected the highest values refer to energy intensive sectors and show an increase from 2000 to 2005 with the noticeable exception of division 34, Motor vehicles.

The most important energy source for all industrial enterprises is by far electricity, followed by natural gas, while the consumption of other products is concentrated on specific activities. As one may see in Graph 3, the use of these two products *per* unit of GVA has increased since the year 2000 while energy saving has occurred for all the other main carriers. This trend has not been hindered by fiscal policy: indeed over these years implicit tax rates generally fell as shown in Section 2.

GRAPH 3

ENERGY INTENSITY BY SOURCE[a] (2000 = 1)

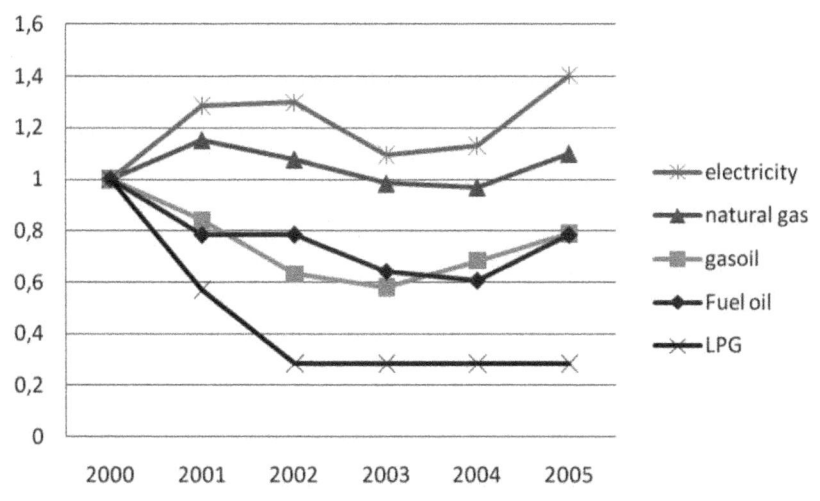

[a] Energy consumption Toe/gross value added in thousands Euro.

In more detail, Graph 4 highlights how for natural gas the incidence of the fiscal component on price (implicit tax rate) is always higher for small and medium enterprises while in the case of electricity the resulting tax rates by classes are more volatile particularly for large firms mainly due to the price variation across the years. This implicit tax rate reduction is mainly due to the fact that the prices for all energy sources have increased since

2000 but if one looks at the cost of energy for firms of different sizes some divergences appear.

GRAPH 4

IMPLICIT TAX RATES BY CLASS OF WORKERS:
NATURAL GAS AND ELECTRICITY (2000 = 100)

Natural gas

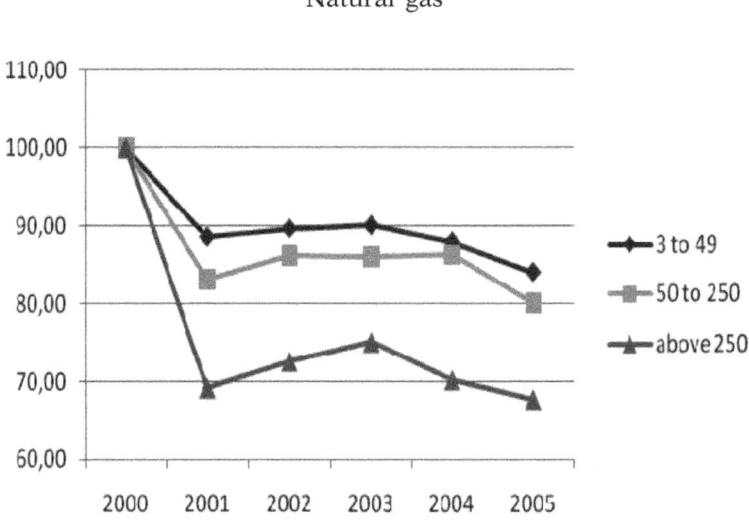

Electricity

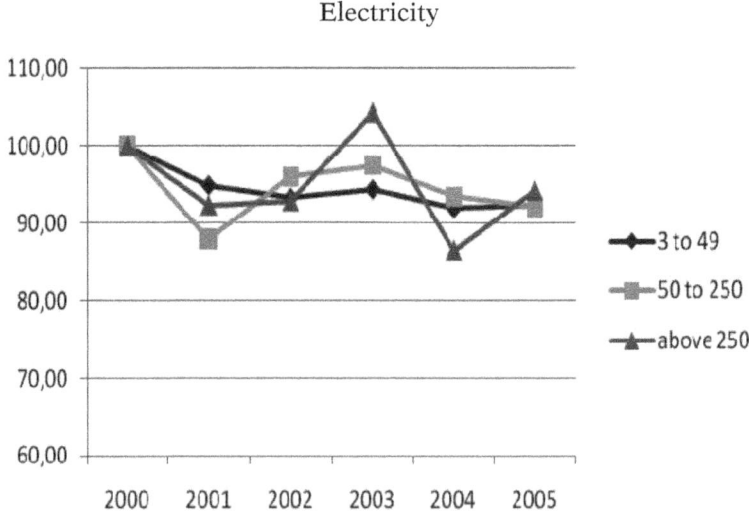

As shown in Table 2 both for natural gas and for electricity, pre-tax prices for small enterprises are always above average and increased in recent years: the opposite trend is shown for firms with more than 250 workers. Indeed, on these markets large enterprises can negotiate lower prices on a special contract basis given their large consumption of energy while small firms are penalized. This considerable cross-sectional price dispersion for both electricity and natural gas is due to a combination of the sector, location and purchase quantity of the firm[30].

TABLE 2

ENERGY PRICES FOR MANUFACTURING FIRMS
IN THE PANEL BY CLASS OF WORKERS (AVERAGE = 100)

	Natural gas			Electricity			Diesel		
	3 to 49	50 to 250	>250	3 to 49	50 to 250	> 250	3 to 49	50 to 250	> 250
2000	105	99	102	105	99	100	105	99	100
2001	104	98	102	103	100	99	98	101	99
2002	109	101	94	108	102	93	109	101	95
2003	110	101	94	108	102	93	109	100	95
2004	112	100	95	104	101	97	107	98	102
2005	107	100	98	103	101	97	112	97	103

As regards the whole set of production factors, Table 3 provides the mean cost shares for our panel. It appears that raw material is the dominating cost with an average share over the period of about 60 per cent while labour is second with a value of about 29 per cent. As one can see from the table the latter trend increased from 2000 onwards as a well-known phenomenon of the Italian economy associating low GDP growth with an increase in employment mostly due to short-term contracts. The shares of capital and energy are lower and fairly stable over time. These shares are slightly different from those of the whole manufactur-

---

[30] This evidence is emphasised in other studies for different countries, such as BJORNER T.B. - TOGEBY M. - JENSEN H.H. (2001) and DAVIS S.J. *et al.* (2008).

ing sector as regards labour and materials (respectively 37 per cent and 51 per cent in 2005) because of the composition bias of our panel, despite their time trend being similar.

TABLE 3

MEANS OF THE SAMPLE COST SHARES

Years	Share of labour	Share of materials	Share of energy	Share of capital
2000	0.275	0.617	0.033	0.075
2001	0.278	0.61	0.034	0.078
2002	0.288	0.602	0.032	0.078
2003	0.292	0.595	0.034	0.079
2004	0.291	0.605	0.031	0.073
2005	0.293	0.602	0.033	0.072
Mean cost share by year, N = 3,575	0.286	0.605	0.033	0.076

6. - Results of the Estimates

The estimated model of this study exploits the advantages of industrial microdata on several issues. This type of data makes it possible to capture the cross-sectional variation in input consumption and to estimate demand elasticities taking into account the within industry between firms variation that is missed by studies using aggregate data. Moreover, using a panel of microdata, we followed up a group of enterprises over time in order to model heterogeneity between firms. Finally, the panel allowed us to compute each firm's average price of labour and energy as described in the previous paragraph. This procedure is the same as that of some previous micro studies (Nguyen and Streitwieser, 2008; Arnberg and Bjorner, 2007) and potentially suffers from an endogeneity problem as these factor prices are likely to reflect differences in the input quality (labour skill mix, energy mix) rather than the exogenous price faced by each firm in the market. Following Arnberg and Bjorner's line of reasoning (2007), we can argue that this problem is limited in panel data models assuming that the input quality within each firm is stable over time and captured by the fixed effect. Moreover, we estimated our model by splitting the sample between small and medium enterprises — with less than 250 workers — and large enterprises, and we introduced dummies for industrial sub-sectors to capture the differences in production technology.

As far as analysis of carbon tax efficacy is concerned, our study is limited by the use of the energy factor as an aggregate to estimate price elasticities, thus failing to take into account the possibility of substitution between energy types. In fact our dataset allows us to construct firms' energy use from several sources but in this study we addressed the issue of factor substitution to investigate the effect of price changes on industrial input mix and assess the potential effect of environmental fiscal policies such as carbon taxation. Moreover, the study of interfuel substitution with microdata poses the problem of zero consumption as it is rare for every single firm to consume all energy types, requiring therefore that an important selection problem should be addressed and corrected.

The system of equations *(5)* was estimated on our micropanel[31] and selected results are presented in Tables 4 and 5 while the complete set of estimated parameters for each production input is listed in the Appendix. Each table presents three sections with estimated factor costs shares and diagnostics, parameters and the implied price elasticities based on the means of estimated factor shares respectively for SMEs and large enterprises. The system was estimated imposing symmetry constraints on factor share equations, while the adding-up and homogeneity restrictions enabled us to derive the parameters for the dropped capital equation[32].

---

[31] The distribution of dependent and independent variables presented some outliers (mostly due to misreporting), which were eliminated from our panel. Therefore the number of observations of our estimates slightly differs from the value in the descriptive statistics of the previous paragraph.

[32] Diagnostic tests were performed to verify the symmetry restrictions and the results confirmed the hypothesis.

The simultaneous estimation of factor cost shares on the basis of relative factor prices, output, individual industrial effects and the technology variable is satisfactory for both groups of firms. In terms of a broad assessment we can say that the results are in line with economic theory and statistically significant (non-statistically significant values at 95 per cent confidence level are shown in bold in the Tables, detailed standard errors are in the Appendix)[33]. Results for SMEs show that each of the four factors is responsive to a change in its own price where η_{LL} = –0.03, η_{MM} = –0.01, and η_{KK} = –0.61. The energy demand is the most elastic with an own-price elasticity η_{EE} = –1.12 in line with similar findings in literature (Agnolucci, 2009). Most of the cross-price elasticities have positive signs with few important negatives such as between energy and capital thus indicating complementarity. At the same time there are substitution possibilities between energy and labour therefore a double dividend effect could take place if ETR is implemented.

Results for large enterprises are more controversial. In fact not all direct price elasticities have the expected sign, with a positive value for labour (η_{LL} = 0.01) and a price elasticity not very different from zero for materials (η_{MM} = 0.002). As far as labour is concerned, aggregation may affect estimated parameters and we can explain the positive price elasticity as large firms tend to hire highly-skilled workers with higher costs according to an efficiency wage argument.

Energy is confirmed as having the most elastic demand to price variation (η_{EE} = –1.22) followed by the demand for capital (η_{KK} = –0.40). As far as energy elasticity is concerned there is no substantial difference between firms classified according to the number of workers. The same conclusion applies if we look at cross-price elasticities: labour is a substitute both of energy (although not statistically significant) and capital even for large firms while energy and capital are complements.

The output parameters (see the Appendix) have a significant

---

[33] Estimated parameters proved to be invariant when the distribution tails were eliminated from the sample, therefore our estimates are robust.

positive influence on all cost shares except energy for large firms
while the time trend used to detect technological change has no
systematic influence on the demand of production inputs. On the
other hand, industrial characteristics play an important role: if we
focus on the energy cost share we can observe that, compared to
the omitted sector Metal ores and other Mining activities, all remaining manufacturing divisions have lower energy requirements
in their input mix with energy intensive activities such as Non-
metallic mineral products being closer to the reference sector. Finally, the number of workers is significant in most cases to indicate a scale effect which operates even within the two broad
groups used here for estimation[34]. In general, our micro-based
analysis is characterized by high heterogeneity which we have
tried to capture by size variables and sectoral dummies that
proved effective[35].

These estimation results give a mixed response in considering the opportunity of an environmental tax reform. On the one
hand the sign and the level of own price elasticity for energy inputs – both for small and large firms – encourages the use of environmental taxes as policy instruments to induce a decrease in
energy intensity; on the other hand cross elasticities show com-

---

[34] In our panel 15 per cent of the sample is represented by firms with less
than 50 workers which behave differently from other firms classified as SMEs.
The same is true for very large enterprises grouped in the class with more than
250 workers.

[35] To test the reliability of SURE estimates and to exploit the potential of our
micro data panel, alternative panel estimates, without constraints and including
the capital equation, have been produced: a Fixed Effect and a Generalized
Equation Estimates. In both cases all variables have been rescaled by using the
root mean square error of a preliminary regression in order to correct for
heteroskedasticity and to obtain a correct comparison with the GLS estimates of
the SURE technique. In the Fixed Effect case it has been possible to estimate the
latent heterogeneity so in this case constants are variable, but the slope coefficients
have the same sign of the SURE estimates notwithstanding a lower significance
for the capital equation. Also in the GE estimate we obtain coefficients with the
same sign and for both of them the complementary relationship between the energy
and capital inputs is confirmed, as well as the substitution between energy and
labour inputs. Moreover, the robustness of the SURE technique has been proved
for each year; in this case it has been possible to obtain a variability of slope and
constant coefficients and we have obtained an elasticity trend with a coherence in
the signs of the estimates. Diagnostic tests on residuals for all equations are good
and can be provided by the authors upon request.

TABLE 4

ESTIMATION RESULTS: SMALL-MEDIUM FIRMS

(a) Factor cost share equations

Equation	Estimated (mean)	Obs	Parms	RMSE	R-sq	chi2
s_lab	0.285	11425	30	0.077	0.755	35263.670
s_mat	0.598	11425	30	0.109	0.663	23056.050
s_ener	0.043	11425	30	0.056	0.201	2914.800

*(b) Coefficients**

Parameters	Labour	Mat.& Serv.	Energy	Capital
Labour	0.194			
Mat.& Serv.	-0.215	0.237		
Energy	0.007	0.007	-0.007	
Capital	0.014	-0.030	-0.007	0.023

(c) Own and cross-price elasticities

Elasticity	Labour	Mat.& Serv.	Energy	Capital
Labour	-0.03	-0.15	0.07	0.12
Mat.& Serv.	-0.07	-0.01	0.05	0.02
Energy	0.45	0.76	-1.12	-0.08
Capital	0.47	0.19	-0.05	-0.61

* As symmetry is imposed, coefficients for any two inputs are shown only once. Parameters and derived elasticities in bold are not significant (below the 95% confidence level).

plementarity between energy and capital which may hamper capital accumulation.

The energy-saving behaviour of firms, however, cannot be the only objective to be pursued. As discussed in Section 2, Italy started with a historically low energy intensity and there is not much room for further improvement in industrial production. On the contrary, the potential for energy input substitution is much greater and the tax instrument permits modulation of tax rates in order to induce the choice of less environmentally-damaging inputs. A first encouraging signal of the role of energy tax rates can be traced in Bardazzi, Oropallo, Pazienza (2009), where the de-

TABLE 5

ESTIMATION RESULTS: LARGE FIRMS

(a) Factor cost share equations

Equation	Estimated (mean)	Obs	Parms	RMSE	R-sq	chi2
s_lab	0.267	4338	29	0.065	0.744	12609.820
s_mat	0.612	4338	29	0.097	0.633	7773.080
s_ener	0.040	4338	29	0.049	0.281	1721.530

*(b) Coefficients**

Parameters	Labour	Mat.& Serv.	Energy	Capital
Labour	0.200			
Mat.& Serv.	-0.214	0.239		
Energy	**0.003**	0.018	-0.010	
Capital	0.012	-0.043	-0.011	0.042

(c) Own and cross-price elasticities

Elasticity	Labour	Mat.& Serv.	Energy	Capital
Labour	0.01	-0.19	0.05	0.12
Mat.& Serv.	-0.08	0.002	0.07	0.01
Energy	0.34	1.07	-1.22	-0.19
Capital	0.41	0.08	-0.09	-0.40

* As symmetry is imposed, coefficients for any two inputs are shown only once. Parameters and elasticities in bold are not significant (below the 95% confidence level).

mands for main energy products (diesel, natural gas, fuel oil and electricity) were estimated by employing a fixed effects model[36]. The input demands were modelled as a function of value added, of the input price (net of taxes), of the specific tax component, of the average price of all other energy inputs used by each firm (excluding the product in question and electricity), of the electricity price, and of sectoral characteristics. Table 6 presents a summary of these estimation results.

[36] The dataset used came from the same source but covered only 2000 and 2004.

TABLE 6

ESTIMATES OF OWN-PRICE AND ENERGY TAX ELASTICITIES*

Own-price elasticities		
	Small enterprises	Large enterprises
Diesel	-0,372	-0,472
natural gas	-1,683	-1,708
fuel oil	**-0,567**	-0,893
electricity	-0,768	**0,182**
Elasticities to Energy taxes		
	Small enterprises	Large enterprises
Diesel	**-0,988**	-3,046
natural gas	-3,233	-2,384
fuel oil	**-0,502**	-2,456
electricity	-0,433	-0,517

* Elasticities in bold are not significant (below the 95% confidence level).
Source: BARDAZZI R. - OROPALLO F. - PAZIENZA M.G. (2009).

For all the estimated equations, all the own-price elasticities (except electricity)[37] have the right (negative) sign and energy product demand is more reactive to changes in tax rates than in net prices, perhaps because changes in taxes are perceived to be more permanent than price changes[38]. These large estimated values confirm that there are opportunities to reduce energy consumption and to modify the input mix by appropriately changing each tax rate.

The energy capital complementarity deserves more attention. Fixed capital is, on average, energy intensive and with the increase of energy prices, the investment choice of firms may be reduced. However, economic literature has shown that for an outcome really improving the environment the quality of capital must change

---

[37] The positive own price elasticity for electricity is not statistically significant. This can be explained by the fact that electricity prices for large firms usually decrease markedly with quantity.

[38] BARKER T. *et* AL. (1995) suggest this interpretation to assess that taxes should give polluters a bigger incentive to reduce CO_2 emissions.

and specific investments in energy saving technology are the key ingredient for a successful path[39]. According to an evolutionary approach[40], a policy signal can drive firms towards new technology and consequently new quality of capital: in other words energy saving technological change is embodied in capital goods. As a consequence, a rise in energy prices may reduce the total amount of capital employed in production but, at the same time, is likely to induce investment and change the quality of capital.

Moreover, estimation of the factor demand system at the sectoral level could deliver interesting insights. Preliminary results on specific energy intensive sectors – such as Non-metallic Minerals, and Pulp, paper, and printing – show positive cross price elasticities between energy and capital thus confirming that the input mix depends on the technological requirements and that the disincentive effect on investments produced by an increase in energy taxes could be restricted to less energy intensive manufacturing sectors and could be tackled by specific industrial policies.

Finally, the possibility of substitution between energy sources needs to be investigated. If ETR is designed with different tax rates across fuels the effects on other production factors will be different. This aspect can be studied by a two step estimation of a sub-energy inputs demand system connected with the full KLEM demand system (Enevoldsen, Anders, Andersen 2007).

7. - Concluding Remarks

After the recent economic crisis, many observers have encouraged a strong link between environmental and industrial policy in order to direct public aid towards environmentally-friendly industrial investments and innovation. Energy saving and emission abatement technologies are particularly attractive in the Ital-

---

[39] Among others, BARDAZZI R. - OROPALLO F. - PAZIENZA M.G. (2009) showed that specific expenditure linked to the emission abatement goal has proved to be significant in the CO_2 emission path of Italian manufacturing firms.

[40] For a review on this issue see CARRARO C. - GERLAGH R. - VAN DER ZWAAN B. (2003) and the recent survey by JAFFE A. - NEWELL R. - POPP D. (2009).

ian context due to the well known delay in meeting Kyoto targets and almost complete energy dependency on imports[41]. However, regulation or aid policies need an enormous amount of information on current research and future innovation paths to be effective, whereas market-based instruments give a signal and leave the choice of investment characteristics to firms, minimizing policy errors and the waste of public resources.

Energy taxation as a market-based instrument for environmental policy can perform an important role if proved to be effective. Our analysis, focused on manufacturing firms, estimated large and negative own price elasticities for energy by employing a translog model estimation on a panel of more than 3000 manufacturing firms for the period 2000-2005. As for other inputs, we found complementarity between energy and capital and substitutability between energy and labour for small firms. It's evident that the encouragement of energy taxes that comes from elasticities' estimation is undermined by the complementarity between energy and capital, which may hamper capital accumulation. However the quality of capital – *i.e.* investment in new technologies – may play a strategic role: a rise in the energy price may cause a reduction of the total amount of capital employed in production but, at the same time, is likely to induce energy-saving investments and a change in the quality of capital. Moreover the price of energy is a weighted average of several energy products employed by firms. Energy taxation, if properly designed, must differentiate tax rates according to the primary goal of policy, giving incentives for a change in the energy input mix and this process, if successful, doesn't necessary lead to a rise in overall energy price. Bardazzi, Oropallo and Pazienza (2009) estimated industrial demand for several energy inputs and found encouraging evidence of substitution among inputs as a consequence of a change in net price and fiscal component.

---

[41] The Italian government has decided to launch a big programme of nuclear power station building. Without entering into a complicated debate on the opportuneness of this strategy, it's important to stress that this process is very slow and benefits for emissions and energy dependency will not be evident for at least fifteen years.

TABLE 7

ESTIMATED PARAMETERS – SMALL AND MEDIUM FIRMS
SHARE OF LABOUR COST

s_lab	Coef.	Std. Err.	z	P>z	95% Conf.	Interval
lnpl	0.194	0.003	56.340	0.000	0.187	0.201
lnpm	-0.215	0.004	-48.640	0.000	-0.223	-0.206
lnpe	0.007	0.002	4.590	0.000	0.004	0.010
lnyR	0.225	0.001	-165.120	0.000	-0.227	-0.222
ladd	0.224	0.002	124.770	0.000	0.220	0.227
_Iyear_2001	0.001	0.002	0.360	0.722	-0.004	0.006
_Iyear_2002	-0.002	0.003	-0.930	0.354	-0.007	0.003
_Iyear_2003	-0.005	0.002	-2.160	0.031	-0.010	0.000
_Iyear_2004	-0.002	0.003	-0.600	0.550	-0.006	0.003
_Iyear_2005	-0.004	0.002	-1.620	0.104	-0.009	0.001
_Iateco2_15	-0.077	0.009	-8.640	0.000	-0.094	-0.059
_Iateco2_17	-0.076	0.009	-8.460	0.000	-0.094	-0.058
_Iateco2_18	-0.012	0.010	-1.220	0.221	-0.031	0.007
_Iateco2_19	-0.079	0.010	-8.100	0.000	-0.099	-0.060
_Iateco2_20	-0.108	0.010	-11.160	0.000	-0.126	-0.089
_Iateco2_21	-0.093	0.010	-9.720	0.000	-0.111	-0.074
_Iateco2_22	0.028	0.010	2.800	0.005	0.008	0.048
_Iateco2_24	-0.065	0.009	-7.220	0.000	-0.083	-0.048
_Iateco2_25	-0.091	0.009	-9.860	0.000	-0.109	-0.073
_Iateco2_26	-0.069	0.009	-7.710	0.000	-0.087	-0.052
_Iateco2_27	-0.067	0.009	-7.350	0.000	-0.085	-0.049
_Iateco2_28	-0.054	0.009	-6.070	0.000	-0.071	-0.037
_Iateco2_29	-0.075	0.009	-8.480	0.000	-0.092	-0.058
Iateco2_30	-0.046	0.016	-2.880	0.004	-0.078	-0.015
_Iateco2_31	-0.077	0.009	-8.190	0.000	-0.096	-0.059
_Iateco2_32	-0.098	0.011	-9.090	0.000	-0.119	-0.077
_Iateco2_33	-0.066	0.010	-6.810	0.000	-0.085	-0.047
_Iateco2_34	-0.084	0.010	-8.810	0.000	-0.103	-0.066
_Iateco2_35	-0.021	0.011	-1.960	0.050	-0.043	0.000
_Iateco2_36	-0.083	0.009	-9.030	0.000	-0.101	-0.065
_cons	2.366	0.018	134.630	0.000	2.332	2.401

TABLE 8

ESTIMATED PARAMETERS – SMALL AND MEDIUM FIRMS
SHARE OF MATERIAL COST

s_mat	Coef.	Std. Err.	z	P>z	95% Conf.	Interval
lnpl	-0.215	0.004	-48.640	0.000	-0.223	-0.206
lnpm	0.237	0.020	11.940	0.000	0.198	0.276
lnpe	0.007	0.002	2.850	0.004	0.002	0.012
lnyR	0.244	0.002	129.050	0.000	0.240	0.248
ladd	-0.238	0.003	-93.770	0.000	-0.243	-0.233
_Iyear_2001	-0.003	0.003	-0.900	0.367	-0.010	0.004
_Iyear_2002	0.002	0.004	0.430	0.665	-0.006	0.009
_Iyear_2003	0.003	0.004	0.810	0.419	-0.004	0.010
_Iyear_2004	0.009	0.004	2.460	0.014	0.002	0.016
_Iyear_2005	0.010	0.004	2.900	0.004	0.003	0.017
_Iateco2_15	0.306	0.013	24.260	0.000	0.282	0.331
_Iateco2_17	0.266	0.013	20.730	0.000	0.241	0.291
_Iateco2_18	0.263	0.014	19.270	0.000	0.236	0.289
_Iateco2_19	0.341	0.014	24.560	0.000	0.314	0.368
_Iateco2_20	0.356	0.014	25.760	0.000	0.329	0.383
_Iateco2_21	0.288	0.013	21.310	0.000	0.261	0.314
_Iateco2_22	0.178	0.014	12.470	0.000	0.150	0.206
_Iateco2_24	0.265	0.013	20.630	0.000	0.239	0.290
_Iateco2_25	0.321	0.013	24.350	0.000	0.295	0.347
_Iateco2_26	0.188	0.013	14.760	0.000	0.163	0.213
_Iateco2_27	0.282	0.013	21.870	0.000	0.257	0.308
_Iateco2_28	0.276	0.013	21.880	0.000	0.251	0.301
_Iateco2_29	0.336	0.013	26.800	0.000	0.312	0.361
_Iateco2_30	0.336	0.023	14.720	0.000	0.291	0.380
_Iateco2_31	0.326	0.013	24.260	0.000	0.300	0.353
_Iateco2_32	0.355	0.016	22.920	0.000	0.325	0.386
_Iateco2_33	0.320	0.014	23.300	0.000	0.293	0.347
_Iateco2_34	0.323	0.014	23.760	0.000	0.296	0.349
_Iateco2_35	0.248	0.015	16.140	0.000	0.218	0.279
_Iateco2_36	0.343	0.013	26.260	0.000	0.317	0.369
_cons	-1.909	0.023	-81.610	0.000	-1.955	-1.863

TABLE 9

ESTIMATED PARAMETERS – SMALL AND MEDIUM FIRMS
SHARE OF ENERGY COST

s_ener	Coef.	Std. Err.	z	P>z	95% Conf.	Interval
lnpl	0.007	0.002	4.590	0.000	0.004	0.010
lnpm	0.007	0.002	2.850	0.004	0.002	0.012
lnpe	-0.007	0.001	-5.170	0.000	-0.010	-0.004
lnyR	-0.003	0.001	-2.950	0.003	-0.005	-0.001
ladd	-0.002	0.001	-1.680	0.093	-0.005	0.000
_Iyear_2001	0.001	0.002	0.290	0.771	-0.003	0.004
_Iyear_2002	0.001	0.002	0.520	0.604	-0.003	0.005
_Iyear_2003	0.001	0.002	0.490	0.623	-0.003	0.004
_Iyear_2004	-0.001	0.002	-0.330	0.739	-0.004	0.003
_Iyear_2005	0.003	0.002	1.550	0.122	-0.001	0.006
_Iateco2_15	-0.137	0.006	-21.260	0.000	-0.150	-0.125
_Iateco2_17	-0.106	0.007	-16.180	0.000	-0.119	-0.093
_Iateco2_18	-0.130	0.007	-18.580	0.000	-0.144	-0.116
_Iateco2_19	-0.148	0.007	-20.790	0.000	-0.162	-0.134
_Iateco2_20	-0.132	0.007	-18.800	0.000	-0.145	-0.118
_Iateco2_21	-0.120	0.007	-17.310	0.000	-0.134	-0.106
_Iateco2_22	-0.143	0.007	-19.600	0.000	-0.157	-0.129
_Iateco2_24	-0.125	0.007	-18.920	0.000	-0.138	-0.112
_Iateco2_25	-0.134	0.007	-19.840	0.000	-0.148	-0.121
_Iateco2_26	-0.052	0.007	-8.000	0.000	-0.065	-0.040
_Iateco2_27	-0.118	0.007	-17.810	0.000	-0.131	-0.105
_Iateco2_28	-0.136	0.006	-21.080	0.000	-0.149	-0.124
_Iateco2_29	-0.155	0.006	-24.150	0.000	-0.167	-0.142
_Iateco2_30	-0.155	0.012	-13.210	0.000	-0.178	-0.132
_Iateco2_31	-0.152	0.007	-22.120	0.000	-0.165	-0.138
_Iateco2_32	-0.147	0.008	-18.800	0.000	-0.162	-0.132
_Iateco2_33	-0.156	0.007	-22.100	0.000	-0.170	-0.142
_Iateco2_34	-0.146	0.007	-21.020	0.000	-0.160	-0.133
_Iateco2_35	-0.142	0.008	-17.980	0.000	-0.158	-0.127
_Iateco2_36	-0.153	0.007	-22.890	0.000	-0.166	-0.140
_cons	0.228	0.011	21.210	0.000	0.207	0.249

TABLE 10

ESTIMATED PARAMETERS – LARGE FIRMS
SHARE OF LABOUR COST

s_lab	Coef.	Std. Err.	z	P>z	95% Conf.	Interval
lnpl	0.200	0.005	40.240	0.000	0.190	0.209
lnpm	-0.214	0.007	-32.670	0.000	-0.227	-0.201
lnpe	0.003	0.002	1.330	0.184	-0.001	0.007
lnyR	-0.211	0.002	-95.850	0.000	-0.215	-0.206
ladd	0.209	0.003	75.820	0.000	0.203	0.214
_Iyear_2001	0.001	0.003	0.230	0.818	-0.006	0.007
_Iyear_2002	-0.003	0.004	-0.950	0.341	-0.010	0.004
_Iyear_2003	-0.004	0.003	-1.170	0.242	-0.011	0.003
_Iyear_2004	0.001	0.003	0.340	0.731	-0.006	0.008
_Iyear_2005	0.000	0.003	-0.060	0.953	-0.007	0.007
_Iateco2_15	-0.145	0.029	-4.980	0.000	-0.202	-0.088
_Iateco2_17	-0.157	0.029	-5.380	0.000	-0.214	-0.100
_Iateco2_18	-0.079	0.029	-2.680	0.007	-0.137	-0.021
_Iateco2_19	-0.153	0.030	-5.120	0.000	-0.212	-0.094
_Iateco2_20	-0.169	0.031	-5.490	0.000	-0.229	-0.109
_Iateco2_21	-0.156	0.030	-5.300	0.000	-0.214	-0.099
_Iateco2_22	-0.025	0.030	-0.850	0.396	-0.083	0.033
_Iateco2_24	-0.147	0.029	-5.030	0.000	-0.204	-0.090
_Iateco2_25	-0.155	0.029	-5.310	0.000	-0.213	-0.098
_Iateco2_26	-0.124	0.029	-4.270	0.000	-0.181	-0.067
_Iateco2_27	-0.166	0.029	-5.680	0.000	-0.223	-0.109
_Iateco2_28	-0.143	0.029	-4.910	0.000	-0.200	-0.086
_Iateco2_29	-0.160	0.029	-5.520	0.000	-0.217	-0.103
_Iateco2_30	(dropped)					
_Iateco2_31	-0.150	0.029	-5.130	0.000	-0.207	-0.093
_Iateco2_32	-0.176	0.030	-5.890	0.000	-0.235	-0.118
_Iateco2_33	-0.147	0.030	-4.930	0.000	-0.205	-0.088
_Iateco2_34	-0.168	0.029	-5.750	0.000	-0.225	-0.111
_Iateco2_35	-0.138	0.030	-4.610	0.000	-0.196	-0.079
_Iateco2_36	-0.144	0.029	-4.910	0.000	-0.202	-0.087
_cons	2.341	0.037	62.890	0.000	2.268	2.414

TABLE 11

ESTIMATED PARAMETERS – LARGE FIRMS
SHARE OF MATERIAL COST

s_mat	Coef.	Std. Err.	z	P>z	95% Conf.	Interval
lnpl	-0.214	0.007	-32.670	0.000	-0.227	-0.201
lnpm	0.239	0.022	10.820	0.000	0.195	0.282
lnpe	0.018	0.003	5.490	0.000	0.012	0.025
lnyR	0.226	0.003	70.330	0.000	0.220	0.232
ladd	-0.225	0.004	-55.300	0.000	-0.233	-0.217
_Iyear_2001	0.000	0.005	-0.070	0.941	-0.010	0.010
_Iyear_2002	0.009	0.005	1.610	0.107	-0.002	0.019
_Iyear_2003	0.009	0.005	1.750	0.080	-0.001	0.019
_Iyear_2004	0.009	0.005	1.760	0.078	-0.001	0.020
_Iyear_2005	0.011	0.005	2.120	0.034	0.001	0.021
_Iateco2_15	0.337	0.044	7.690	0.000	0.251	0.423
_Iateco2_17	0.288	0.044	6.550	0.000	0.202	0.374
_Iateco2_18	0.286	0.044	6.470	0.000	0.200	0.373
_Iateco2_19	0.362	0.045	8.030	0.000	0.274	0.450
_Iateco2_20	0.370	0.046	7.990	0.000	0.280	0.461
_Iateco2_21	0.270	0.044	6.060	0.000	0.182	0.357
_Iateco2_22	0.217	0.044	4.890	0.000	0.130	0.304
_Iateco2_24	0.294	0.044	6.710	0.000	0.208	0.380
_Iateco2_25	0.342	0.044	7.780	0.000	0.256	0.429
_Iateco2_26	0.182	0.044	4.150	0.000	0.096	0.268
_Iateco2_27	0.325	0.044	7.390	0.000	0.239	0.411
_Iateco2_28	0.313	0.044	7.140	0.000	0.227	0.399
_Iateco2_29	0.373	0.044	8.530	0.000	0.287	0.458
_Iateco2_30	(dropped)					
_Iateco2_31	0.354	0.044	8.030	0.000	0.267	0.440
_Iateco2_32	0.397	0.045	8.770	0.000	0.308	0.485
_Iateco2_33	0.340	0.045	7.600	0.000	0.252	0.428
_Iateco2_34	0.358	0.044	8.140	0.000	0.272	0.444
_Iateco2_35	0.340	0.045	7.570	0.000	0.252	0.428
_Iateco2_36	0.365	0.044	8.260	0.000	0.278	0.451
_cons	-1.764	0.055	-32.310	0.000	-1.871	-1.657

TABLE 12

ESTIMATED PARAMETERS – LARGE FIRMS
SHARE OF ENERGY COST

s_mat	Coef.	Std. Err.	z	P>z	95% Conf.	Interval
lnpl	0.003	0.002	1.330	0.184	-0.001	0.007
lnpm	0.018	0.003	5.490	0.000	0.012	0.025
lnpe	-0.010	0.002	-5.650	0.000	-0.014	-0.007
lnyR	-0.002	0.002	-1.050	0.293	-0.005	0.001
ladd	0.003	0.002	1.500	0.134	-0.001	0.007
_Iyear_2001	0.002	0.003	0.600	0.548	-0.003	0.007
_Iyear_2002	-0.003	0.003	-1.230	0.217	-0.008	0.002
_Iyear_2003	-0.001	0.003	-0.580	0.559	-0.007	0.004
_Iyear_2004	-0.002	0.003	-0.880	0.376	-0.007	0.003
_Iyear_2005	0.001	0.003	0.290	0.769	-0.004	0.006
_Iateco2_15	-0.164	0.022	-7.450	0.000	-0.207	-0.121
_Iateco2_17	-0.133	0.022	-6.040	0.000	-0.176	-0.090
_Iateco2_18	-0.169	0.022	-7.610	0.000	-0.212	-0.125
_Iateco2_19	-0.175	0.023	-7.770	0.000	-0.220	-0.131
_Iateco2_20	-0.156	0.023	-6.750	0.000	-0.202	-0.111
_Iateco2_21	-0.123	0.022	-5.550	0.000	-0.167	-0.080
_Iateco2_22	-0.183	0.022	-8.240	0.000	-0.226	-0.139
_Iateco2_24	-0.144	0.022	-6.560	0.000	-0.187	-0.101
_Iateco2_25	-0.166	0.022	-7.520	0.000	-0.209	-0.123
_Iateco2_26	-0.076	0.022	-3.460	0.001	-0.119	-0.033
_Iateco2_27	-0.130	0.022	-5.910	0.000	-0.173	-0.087
_Iateco2_28	-0.160	0.022	-7.290	0.000	-0.203	-0.117
_Iateco2_29	-0.179	0.022	-8.170	0.000	-0.222	-0.136
_Iateco2_30	(dropped)					
_Iateco2_31	-0.177	0.022	-8.030	0.000	-0.220	-0.134
_Iateco2_32	-0.184	0.023	-8.180	0.000	-0.229	-0.140
_Iateco2_33	-0.180	0.022	-8.060	0.000	-0.224	-0.136
_Iateco2_34	-0.167	0.022	-7.610	0.000	-0.210	-0.124
_Iateco2_35	-0.185	0.022	-8.240	0.000	-0.229	-0.141
_Iateco2_36	-0.179	0.022	-8.100	0.000	-0.222	-0.136
_cons	0.202	0.026	7.820	0.000	0.152	0.253

BIBLIOGRAPHY

AGNOLUCCI P., «Ex Post Evaluation of CO_2 Taxes: A Survey», *Working Paper,* no. 52, Tyndall Centre, 2004.

— - —, «The Effect of the German and British Environmental Taxation Reforms: A Simple Assessment», *Energy Policy,* no. 37, 2009, pages 3043-3051.

ARNBERG S. - BJORNER T.B., «Substitution between Energy, Capital and Labour within Industrial Companies: A Micro Panel Data Analysis», *Resource and Energy Economics,* no. 29, 2009, pages 122-136.

BARDAZZI R. - OROPALLO F. - PAZIENZA M.G., «Accise energetiche e competitività delle imprese: un'applicazione sull'esperimento della carbon tax» (Energy Taxes and Industrial Competitiveness: The Case of Italian Carbon Tax), *Economia delle fonti di energia e dell'ambiente,* no. 3, 2004, pages 121-164.

— - —, — - —, — - —, «Complying Kyoto Targets: An Assessment of Energy Taxes Effectiveness in Italy», in ZAIDI A. - HARDING A. - WILLIAMSON P. (eds.), *New Frontiers in Microsimulation Modelling,* Ashgate, 2009, pages 605-629.

BARDAZZI R. - PARISI V. - PAZIENZA M.G., «Modelling Direct and Indirect Taxes on Firms: A Policy Simulation», *Austrian Journal of Statistics,* Vol. 33 (1+2), 2004, pages 237-259.

BARKER T. - EKINS P. - JOHNSTONE N., *Global Warming and Energy Demand,* Routledge, London, 1995.

BJORNER T.B. - TOGEBY M. - JENSEN H.H., «Industrial Companies' Demand for Electricity: Evidence from a Micropanel», *Energy Economics,* no. 23, 2001, pages 595-617.

CARRARO C. - GERLAGH R. - VAN DER ZWAAN B., «Endogenous Technical Change in Environmental Macroeconomics», *Resource and Energy Economics,* no. 25, 2003, pages 1-10.

CHRISTENSEN L.R. - JORGENSON D.W. - LAU L.J., «Transcendental Logarithmic Production Frontiers», *Review of Economics and Statistics,* no. 55, 1973, pages 28-45.

— - —, — - —, — - —, «Transcendental Logarithmic Utility Functions», *American Economic Review,* no. 65(3), 1975, pages 367-383.

DAVIS S.J. - GRIM C. - HALTIWANGER J. - STRITWIESER M., «Electricity Pricing to US: Manufacturing Plants, 1963-2000», *NBER, Working Paper,* no. 13778, 2008.

EKINS R. - SPECK S., «Competitiveness and Exemptions from Environmental Taxes in Europe», *Environmental and Resource Economics,* no. 13, 1999, pages 369-396.

ENEVOLDSEN M.K. - ANDERS V.R. - ANDERSEN M.S., «Decoupling of Industrial Consumption and CO_2 Emissions in Energy-Intensive Industries in Scandinavia», *Energy Economics,* no. 29, 2007, pages 665-692.

EUROPEAN COMMISSION, *Draft Proposal for a Council Directive Amending Directive 2003/96 Restructuring the Community Framework for the Taxation of Energy Products and Electricity,* Bruxelles, 2009.

— - —, *Excise Duty Tables Part II - Energy Products and Electricity,* Directorate general taxation and customs union, 2010.

JAFFE A. - NEWELL R. - POPP D., «Energy, the Environment, and Technological Change», *NBER, Working Paper,* no. 14832, 2009.

LAURENT E. - LE CACHEX J. «An Ever Less Carbonated Union? Towards a Better European Taxation Against Climate Change», *Notre Europe Studies and Re-*

searches, no. 74, 2009.

LIU G., «Estimating Energy Demand Elasticities for OECD Countries: A Dynamic Panel Data Approach», *Discussion Papers*, no. 373, Statistics Norway, Research Department, 2004.

KOETSE M.J. - DE GROOT H.L.F. - FLORAX R.J.G.M., «Capital-Energy Substitution and Shifts in Factor Demand: A Meta-Analysis», *Energy Economics*, no. 30, 2008, pages 2236-2251.

MEDINA J. - VEGA-CERVERA J.A., «Energy and the Non-Energy Inputs Substitution: Evidence for Italy, Portugal and Spain», *Applied Energy*, no. 68, 2001, pages 203-214.

NGUYEN S.V. - STREITWIESER M.L. , «Capital-Energy Substitution Revisited: New Evidence from Micro Data», *Journal of Economic and Social Measurement*, no. 33, 2008, pages 129-153.

OECD, *The Political Economy of Environmentally Related Taxes*, OECD, Paris, 2006.

PEARSON M., «The Political Economy of Implementing Environmental Taxes», *International Tax and Public Finance*, Vol. 2, no. 2, 1995, pages 357-373.

SPECK S., «Overview of Environmental Tax Reform in EU Member States», Cometr Final Report to the European Commission, *http://www2.dmu.dk/cometr /COMETR_Final_Report.pdf*, 2007.

SOLOW J.L., «The Capital-Energy Complementarity Debate Revisited», *The American Economic Review*, Vol. 77, no. 4, 1987, pages 605-614.

THOMPSON H., «The Applied Theory of Energy Substitution in Production», *Energy Economics*, no. 28, 2006, pages 410-425.

ZELLNER A., «An Efficient Method of Estimating Seemingly Unrelated Regression Equations and Tests for Aggregation Bias», *Journal of the American Statistical Association*, no. 57, 1962, pages 348-368.

Financing Public Expenditure *via* Emissions Taxation under International Emissions Trading: Is There Any Scope for Emission Tax Harmonization?

Alessio D'Amato - Amanda Spisto*

University of Rome "Tor Vergata"

We address the issue of emission tax harmonization in a model featuring two representative firms located in two countries. Firms are subject to an international emissions trading system and to domestic emissions taxation; the latter generates public revenue but also implies implementation costs. Decentralized tax setting causes a spillover across countries via the permits price. Nonetheless, harmonization might imply a lower aggregate social welfare. This happens when uniform taxation prevents the exploitation of significant differences across countries in terms of costs and benefits of taxation. Finally, we identify cases where harmonization implies larger aggregate social welfare but lacks unanimous consent. [JEL numbers: Q58, H23].

Keywords: environmental tax harmonization; international emissions trading; transboundary pollution.

1. - Introduction

As climate policy develops, and as the deadlines of interna-

---

* <*damato@economia.uniroma2.it*>, <*spisto@economia.uniroma2.it*>, Dipartimento SEFEMEQ, Facoltà di Economia, Università degli Studi di Roma "Tor Vergata". The Authors would like to thank Bouwe Dijkstra for insightful comments and suggestions. Errors are, of course, Authors' responsibility.

tional environmental agreements get closer, a growing attention is devoted to instruments that might complement current policies, in the EU as well as worldwide. The EU case is particularly interesting, as a GHG emissions trading system has to coexist with national environmental policies aimed at objectives which might differ from GHG emission reduction. The mix of national and supranational instruments results in complementarities, but there is a risk of overlapping and conflicts. For example, a survey on the potential interaction of the UK climate policy with the European ETS is presented by Sorrell and Sijm (2003), who argue that the EU ETS is incompatible with the climate policy in the UK due to distributional effects upon different groups, double regulation, double counting, as well as differential treatment of regulated and non regulated sectors.

The problem of overlapping regulation is particularly important when a country that becomes part of an international emissions trading system also implements taxation of environment/energy intensive goods (for example fuels for transport or heating) in order to raise revenue to finance public expenditures. This is likely to be an issue because the setting of national taxes, given the overall target each country has to achieve in terms of emissions, can generate spillovers among countries due to the presence of the international allowance market. In such a case, the harmonization of tax rates could be a way out from the resulting losses in social welfare. However, such harmonization can be difficult to implement, the pressure against a EU wide carbon tax being a real life example (Pearce, 2005).

The aim of our paper is twofold:

– we assess the spillovers among countries arising when revenue raising domestic taxes on emissions overlap with an international emissions trading system (*fiscal externalities* in the terminology of Eichner and Pethig, 2009);

– we move a first step in the investigation of the consequences of such spillovers on social welfare and in the evaluation of tax harmonization as a suitable remedy when benefits and costs from taxation are explicitly accounted for.

We model a two countries setting with two representative firms (one in each country). Firms are subject to two kinds of regulation: an international emissions trading system, where they act as price takers, and domestic emission taxation aimed at raising public revenue to finance socially beneficial public expenditure. Countries are assumed to be symmetric except with respect to the marginal benefits of public expenditure and the related implementation costs. Under no harmonization, each country's government chooses the domestic tax rate non cooperatively, while in a harmonized framework the tax rate is (uniform across countries and) chosen by a supranational regulator. Finally, the emissions target is assumed as exogenous (coherently with existing international environmental negotiations).

Two effects determine the results of our paper: on one hand, an increase in the tax rate in any country implies a lower production, *i.e.* lower emissions in that country, and *ceteris paribus* a lower permits price and a higher production in the other country. Such tax related spillover implies a lower social welfare if the tax rates are not harmonized across countries. On the other hand, differences in marginal benefits and costs of public expenditure cannot be properly "exploited" in a harmonized setting, for example by raising more public revenue where it is more beneficial and/or less costly to do so.

Our analysis leads us to the following conclusions. As already anticipated we first identify a spillover across countries vehiculated by the permits price and related to decentralized (*i.e.* not harmonized) tax setting. Such spillover implies that harmonization is optimal if countries are completely symmetric. Then we introduce asymmetries in the costs and benefits of taxation and show that a larger differentiation leads to a higher likelihood that social welfare is lower under harmonization. This is due to the fact that, when benefits and costs from public expenditure are similar across countries, the tax related spillover dominates in terms of the impact on aggregate social welfare, while when taxation implies substantially lower costs and significantly larger benefits in one country, then harmonization is not good for aggregate welfare, as it imposes the same tax rates to all countries. Finally, and interestingly, we identify cases where at least one country might not consent to harmonizing tax rates even when it would be aggregate welfare improving to do so. The latter result could be relevant when there is no politically feasible way of compensating the losing countries and the introduction of fiscal measures requires unanimity (as it is the case in the EU).

Our paper is closely related to Bohringer *et* al. (2008), who address the consequences of the unilateral introduction of carbon emissions taxation in sectors which are already subject to an international emissions trading system. The authors point out that emission taxes overlapping with international emissions trading are in general ineffective and costly, unless emission taxes themselves are levied in a sufficiently large number of member countries and at a high enough rate. Along the same lines, Johnstone (2003) underlines that the use of a mix of policy instruments to achieve one objective «... will be at best redundant and at worst counterproductive». An interesting implication of results in Bohringer *et* al. (2008), also obtained in the general equilibrium setting developed by Eichner and Pethig (2009), is that efficiency is not damaged by overlapping regulation if tax rates are uniform across countries which are part of the international emissions trading scheme.

Other papers, dealing with environmental policy harmonization, are also related to our work. Ulph (1997) analyses the case of harmonization and optimal environmental policy in a federal system with asymmetric information and considers the case in which states differ from the point of view of key characteristics such as technologies for abating pollution, natural resource endowments, *etc*. The author advocates the need for a "flexible" policy harmonization, accounting for environmental differences across countries. Agostini *et* al. (1992) analyse the effect of a carbon tax to reduce CO_2 emissions in Europe and conclude that country specific environmental policies should be adopted; moreover, international coordination should be concerned with environmental targets rather than with environmental instruments. Gusdorf (2007) studies the stability of international fiscal harmonization in the case of climate change policies. He argues that heterogeneity concerns among countries subject to an emission tax harmonization policy would bring the system to instability; he concludes the analysis considering that intra- and inter-coalition financial transfers are necessary – but not sufficient – to stabilize harmonization. Finally, our paper is also somewhat linked to the literature on the impact of non-environmental policies on the environment, very well exemplified in Markandya (2005).

We add to the received literature by explicitly allowing for benefits and costs related to environmental taxation (which might differ across countries) and by showing that the inclusion of such benefits and costs might lead nonuniform tax rates to be welfare improving with respect to uniform ones when national revenue raising taxes and international emissions trading overlap.

The paper is organized as follows: the next section presents the model and solves the firms' problem, while section 3 analyses the regulatory framework arising under no harmonization. Section 4 derives results corresponding to the harmonization case and section 5 performs policy relevant comparisons among the two institutional settings. Finally, section 6 concludes.

2. - The Model and the Firms' Problem

We assume there are two countries (i and j), each featuring a large set of identical firms. We deal, therefore, with two representative firms, one from each country. Both firms are assumed to be subject, at the same time, to an international emissions trading system and to domestic emissions taxation, the latter intended to raise revenue to finance public expenditure. To keep matters simple, we assume a 1 to 1 relationship between production and polluting emissions.

The objective function of firm in country $k = i, j$ is:

(1)
$$\Pi_k = \alpha q_k - \frac{\beta}{2} q_k^2 - t_k q_k - p(q_k - e_k)$$

where q_k are polluting emissions (and production), t_k are unit emission taxes, p is the price of emission permits and, finally, e_k is the amount of allowances issued in country $k = i, j$. We assume that the amount of emission allowances issued in each country is exogenous, *i.e.* out of control of national and/or supranational environmental regulator(s). In other words, the option of harmonizing (or not) taxation is considered after the pollution reduction targets as well as the "burden sharing" among countries to achieve such targets have been chosen[1]. Also, we assume that the auctioning of permits is very difficult to be implemented due to political pressures from powerful regulated sectors, so that permits are allocated for free[2].

The permits market is perfectly competitive, *i.e.* firms are price takers. Notice that the only potential source of asymmetries across firms in different countries is related to the tax rate. The implications of this assumption will be discussed shortly.

The first order (necessary and sufficient) conditions for an optimum are as follows:

$$(2) \qquad\qquad \alpha - \beta q_k - t_k - p = 0$$

so that the equilibrium level of emissions for any level of emission taxes and permits price is:

$$(3) \qquad\qquad q_k = \frac{1}{\beta}\left(\alpha - p - t_k\right)$$

Equilibrium on the permits market requires that total emissions equal the total number of issued allowances, that we label as $E = e_i + e_j$:

$$q_i + q_j = E$$

---

[1] This is, in our view, close to actuality. For example, in the largest existing emissions trading system, the EU ETS, the amount of permits to be issued in each country (as well as at the EU level) has to be coherent with existing environmental agreements among member countries (the so-called "burden sharing" Directive) as well as among the EU and other countries participating to climate change international negotiations.

[2] Although the latest changes to the EU ETS (Directive 2009/29/CE) explicitly introduce the auctioning of emission permits, the role of auctioning has been, so far, limited, due to political implementation problems (see, among others, MARKUSSEN P. - SVENDSEN G.T., 2005).

leading to the following equilibrium price of permits:

(4)
$$p = \alpha - \frac{1}{2}(t_i + t_j) - \frac{1}{2}\beta E$$

Notice that the tax rates affect the equilibrium permits price. More specifically, an increase in t_k ($k = i, j$) implies *ceteris paribus* a less than proportional decrease in the equilibrium price of emission allowances, *i.e.* $\dfrac{\partial p}{\partial t_k} = -\dfrac{1}{2}$ for all $k = i, j$.

Equilibrium production/emission levels are as follows:

(5)
$$q_i = \frac{1}{2}\left(\frac{t_j - t_i}{\beta} + E \right)$$

(6)
$$q_j = \frac{1}{2}\left(\frac{t_i - t_j}{\beta} + E \right)$$

From *(4)* we can expect the choice of the tax rate in one country to affect welfare in the other country *via* the emission permits market. This is somewhat confirmed if we focus on the impact of taxation on production/emissions. Indeed, in equilibrium:

$$\frac{\partial q_i}{\partial t_i} = -\frac{1}{\beta}\left(\frac{\partial p}{\partial t_i} + 1 \right) = -\frac{1}{2\beta}$$

$$\frac{\partial q_j}{\partial t_i} = \frac{1}{2\beta}$$

As a consequence, an increase in the tax rate in country i leads to an increase in production and emissions in country j *via* the permits' price. Of course, given the overall environmental target E, such production changes match exactly in equilibrium. On the other hand, a welfare impact can take place due to the "distribution" of emissions across countries.

3. - The Regulatory Problem: No Harmonization

Moving to the objective function of the regulator(s), under no harmonization each country chooses its own tax rate non cooperatively, to maximize domestic social welfare. The latter equals the firms' profit function from production/emissions (including firms' profits and expenses (revenues) from buying (selling) permits), *minus* damages from pollution *plus* social benefits from public expenditure *minus* implementation and administrative costs related to taxation[3]

(7) $$ W_k = \alpha q_k - \frac{\beta}{2} q_k^2 - p(q_k - e_k) - \frac{d}{2} E^2 + \mu_k t_k q_k - \frac{\lambda_k}{2} t_k^2 $$

where:

$-\dfrac{d}{2} E^2$ are damages from pollution, which are assumed to be striclty convex in emissions (d is a positive parameter); we assume transboundary pollution takes place, in order for the assumption of international emissions trading to be meaningful. Of course, as the overall emission target is exogenous, this terms can be considered here as a constant.

$- \mu_k t_k q_k$ are the benefits stemming from public expenditure financed *via* emissions taxation, which are assumed to depend linearly on total revenue $t_k q_k$ by a positive parameter μ_k.

The last term of (7) deserves more comments. As Smulders and Vollebergh (2001) underline, environmental taxes are subject to administrative and implementation costs that are likely to vary with the tax rates, a higher rate implying higher costs. For example, a higher tax rate could imply larger incentives towards tax evasion and, at the same time, larger monitoring and compliance costs. Further, we can expect that a larger existing tax rate will generate a larger resistance by regulated sectors to a further marginal increase in the tax rate itself. As a result, and in the absence of conclusive empirical evidence, we simply assume that a larger

---

[3] Consumer's surplus is not explicitly included in the objective function (7), as total output and the related price are given.

existing tax rate requires larger marginal costs to increase it further, *i.e.* implementation costs are convex in the tax rate. In *(7)* λ_k is a positive exogenous parameter determining the degree of convexity of implementation costs.

The two countries are assumed to be "almost" symmetric, in that they only differ in terms of the marginal benefits from public expenditure, *i.e.* $\mu_i \neq \mu_j$ and the implementation cost parameters, *i.e.* $\lambda_i \neq \lambda_j$.

First order conditions with respect to country i imply

$$t_i = \frac{1}{4\mu_i + 4\beta\lambda_i + 3}\left(\left(2\mu_i + 1\right)t_j + \beta\left(e_j - e_i + 2\mu_i e_i + 2\mu_i e_j\right)\right)$$

while, following the same reasoning for country j we get:

$$t_j = \frac{1}{4\mu_j + 4\beta\lambda_j + 3}\left(\left(2\mu_j + 1\right)t_i + \beta\left(e_i - e_j + 2\mu_j e_i + 2\mu_j e_j\right)\right)$$

Notice that emissions tax rate in country i increases with the corresponding tax rate in country j and *vice versa*.

The solution to the system implies that the equilibrium tax rates are as follows:

$$t_i = 2\Omega\lambda_j\beta\left(e_j - e_i + 2\mu_i\left(e_i + e_j\right)\right) +$$
$$(8) \qquad + \Omega\left(e_j - e_i + 4\mu_i e_i + 2\mu_i e_j - \mu_j e_i + 3\mu_j e_j + 6\mu_i\mu_j\left(e_i + e_j\right)\right)q$$

and

$$t_j = 2\Omega\lambda_i\beta\left(e_i - e_j + 2\mu_j\left(e_i + e_j\right)\right) +$$
$$(9) \qquad + \Omega\left(e_i - e_j + 3\mu_i e_i - \mu_i e_j + 2\mu_j e_i + 4\mu_j e_j + 6\mu_i\mu_j\left(e_i + e_j\right)\right)$$

where

$$\Omega = \frac{\beta}{5\mu_i + 5\mu_j + 6\beta\lambda_i + 6\beta\lambda_j + 6\beta\mu_i\mu_j + 8\beta^2\lambda_i\lambda_j + 8\beta\lambda_i\mu_j + 8\beta\lambda_j\mu_i + 4}$$

The corresponding equilibrium emissions and price levels can be obtained by substituting from *(8)* and *(9)* in *(5)*, *(6)* and *(4)*.

We label optimal social welfare under no harmonization in

country i and j respectively as W_i^n and W_j^n, while $W^n = W_i^n + W_j^n$ is the corresponding aggregate welfare.

4. - The Regulatory Problem: Harmonization

This section is devoted to the analysis of the harmonized case. We assume here that the tax rate is chosen by a centralized regulator maximizing the sum of involved countries' social welfare. Coherently with the no harmonization case, we assume that the total amount of issued emission allowances is exogenous (for example set by international negotiations). The corresponding social welfare function is:

$$(10) \qquad W = \alpha E - \beta q^2 - dE^2 + tq\left(\mu_i + \mu_j\right) - \left(\frac{\lambda_i}{2} + \frac{\lambda_j}{2}\right)t^2$$

where t is the harmonized tax rate, while q is the corresponding output level[4]. The choice of t is constrained by the reaction of firms which, from *(3)*, can be written as follows:

$$(11) \qquad q = \frac{1}{\beta}\left(\alpha - x - t\right)$$

where we label x as the equilibrium permits price in the harmonization case. Equilibrium on the permits market requires:

$$2q = E$$

implying:

$$(12) \qquad x = \alpha - t - \frac{1}{2}\beta E$$

The corresponding equilibrium tax rate and output/emission levels are as follows:

---

[4] Notice that output levels in a harmonized setting are identical across countries. See condition *(11)* below.

(13)
$$t = \frac{1}{2}\frac{\mu_i + \mu_j}{\lambda_i + \lambda_j}\left(e_i + e_j\right)$$

(14)
$$q = \frac{1}{2}\left(e_i + e_j\right)$$

We label W_i^h and W_j^h optimal social welfare under harmonization in country i and j respectively. Also, we label W^h as the corresponding aggregate social welfare.

5. - Comparisons

5.1. *Identifying the Spillovers*

The aim of this section is to perform comparisons between the harmonized and the no harmonization case. Two opposing forces are expected to drive our results:
– on one hand, the presence of relevant spillovers across countries, which are vehiculated by the permits price[5]. More specifically, an increase in the tax rate in one country implies a decrease in p and, as a consequence, an increase in production/emissions in the other country. Though overall production is constant (as $q_i + q_j = e_i + e_j = E$), the distribution of emissions across countries might influence public revenue and social welfare. To identify the spillovers among countries, we focus on country i. Differentiating *(7)* with $k = i$ with respect to t_j and accounting for *(2)* and *(4)* we get:

$$\frac{\partial W_i}{\partial t_j} = \left(\mu_i + 1\right)t_i\frac{\partial q_i}{\partial t_j} - \frac{\partial p}{\partial t_j}\left(q_i - e_i\right)$$

As, from *(5)*, q_i increases with t_j, i.e. $\frac{\partial q_i}{\partial t_j} - \frac{1}{2\beta}$ we can rewrite the "spillover equation" as follows:

---

[5] The spillover identified in this section is coherent with what EICHNER T. - PETHIG R. (2009) call *fiscal externalities*.

(15)
$$\frac{\partial W_i}{\partial t_j} = \frac{1}{2\beta}(\mu_i + 1)t_i + \frac{1}{2}(q_i - e_i)$$

The first term in *(15)* is positive, while the second is positive (negative) when country i is a net buyer (seller) of permits. The net sign of $\frac{\partial W_i}{\partial t_j}$ cannot be determined *a priori*. In any case, the presence of spillovers implies, *ceteris paribus*, a higher social welfare under harmonization;

– on the other hand, harmonization does not allow the central government to account for countries specificities in terms of marginal benefits of public expenditures and/or in terms of implementation costs; this is likely to decrease social welfare under harmonization.

5.2. Complete Symmetry

As a benchmark for our comparisons, we start by addressing a case where countries are completely symmetric. Although such case is trivial under a social welfare point of view, it is useful to investigate the nature of spillovers vehiculated by the international permits market.

First of all, as we are not interested in the consequences of varying (exogenous) initial allocation of permits, we normalize E to 1, and assume $e_i = e_j = \frac{1}{2}E = \frac{1}{2}$ *i.e.* complete symmetry in endowments. Also, $\mu_i = \mu_j = m$ and $\lambda_i = \lambda_j = l$. It is easily shown that in such a case $q_i = q_j = q = \frac{1}{2}$. Further, under no harmonization we get[6]:

$$t_i = t_j = \beta \frac{m}{m + 2l\beta + 1}$$

---

[6] Notice that, given complete symmetry, tax rates are identical across countries even in the absence of "explicit" harmonization, *i.e.* in the no harmonization case.

so that

$$p = \alpha - \frac{1}{2}\frac{\beta(3m + 2l\beta + 1)}{m + 2l\beta + 1},$$

while, in the "harmonized taxes" case,

$$t = \frac{m}{2l}$$

so that

$$x = \alpha - \frac{1}{2}\frac{m + l\beta}{l}.$$

We can therefore compare equilibrium tax rates and permits prices:

$$t_i - t = t_j - t = -\frac{1}{2}m\frac{m + 1}{l(m + 2l\beta + 1)} < 0$$

This is reasonable. Under perfect symmetry, $q_k = e_k$ for all k; as a result, the spillover among countries under no harmonization is positive. In the latter setting, therefore, countries will set tax rates which are lower than the socially optimal ones. The resulting public revenue will be larger under harmonization, while equilibrium permits price will be larger under no harmonization. Indeed,

$$p - x = \frac{1}{2}m\frac{m + 1}{l(m + 2l\beta + 1)} > 0$$

Finally, as expected, social welfare is higher under harmonized taxes, as (substituing for equilibrium values in *(7)* and *(10)*):

$$W^h - (W_i^n + W_j^n) = \frac{1}{4}m^2\frac{(m + 1)^2}{l(m + 2l\beta + 1)^2}$$

We get therefore the first result of our paper.

PROPOSITION 1. *When we assume perfectly symmetric countries, tax rates and public revenue are lower under no harmonization. The corresponding equilibrium permits price is higher.*

The result summed up in Proposition 1 is consistent with the existing literature and is a direct consequence of the previosuly identified spillover across countries when tax rates are set non co-operatively. Under the symmetry assumption, such spillover is not counterbalanced by any cost or benefit differential across countries, so that social welfare is strictly higher under harmonization.

5.3. *Introducing Asymmetries across Countries*

The aim of this section is to investigate how our results change if we introduce asymmetries across countries in terms of implementation costs due to taxation and/or marginal benefits from public expenditure. In order to investigate the net effects on emissions, tax rates and social welfare, we focus on specific parameter values, summed up in Table 1[7].

TABLE 1

PARAMETER VALUES

$\mu_i = 1$	$\mu_j = \theta$	$\lambda_i = 1$	$\lambda_j = \sigma$	$\alpha = 10$

We will keep in what follows the assumptions $E = 1$ and $e_i = e_j = \frac{1}{2}E$.

Parameters θ and σ represent the degree of asymmetry across countries. For $\theta \in (0,1)$ marginal benefits are larger in country i, while when $\theta > 1$ the opposite holds. $\sigma \in (0,1)$ implies a larger implementation cost parameter in country i, while the opposite holds when a $\sigma > 1$. Under the assumed parameter values, the

---

[7] Of course, the range for β as well as the assumed parameter values will be such to guarantee that all variables are strictly positive. Also, the allowed range of variation of θ and σ is not too wide (between 0.5 and 2) in order to avoid "extreme" scenarios and to better underline our conclusions, without affecting the bulk of our results.

spillover arising under no harmonization can be shown to be positive, and the following levels for emissions, tax levels and equilibrium prices are obtained:

TABLE 2

EQUILIBRIUM EMISSIONS, TAX RATES AND PERMITS PRICES

Variable	No harmonization	Harmonization
emiss. i	$q_i = \dfrac{1}{2}\dfrac{13\theta + 6\beta + 12\theta\beta + 10\sigma\beta + 8\sigma\beta^2 + 7}{11\theta + 6\beta + 8\theta\beta + 14\sigma\beta + 8\sigma\beta^2 + 9}$	$q = \dfrac{1}{2}$
emiss. j	$q_j = \dfrac{1}{2}\dfrac{9\theta + 6\beta + 4\theta\beta + 18\sigma\beta + 8\sigma\beta^2 + 11}{11\theta + 6\beta + 8\theta\beta + 14\sigma\beta + 8\sigma\beta^2 + 9}$	$q = \dfrac{1}{2}$
tax rate i	$t_i = \beta\dfrac{7\theta + 4\sigma\beta + 3}{11\theta + 6\beta + 8\theta\beta + 14\sigma\beta + 8\sigma\beta^2 + 9}$	$t = \dfrac{1}{2}\dfrac{\theta+1}{\sigma+1}$
tax rate j	$t_j = \beta\dfrac{9\theta + 4\theta\beta + 1}{11\theta + 6\beta + 8\theta\beta + 14\sigma\beta + 8\sigma\beta^2 + 9}$	$t = \dfrac{1}{2}\dfrac{\theta+1}{\sigma+1}$
eq. price	$p = \dfrac{1}{2}\dfrac{22\theta + 107\beta - 6\beta^2 + 133\theta\beta + 280\sigma\beta - 12\theta\beta^2 + 142\sigma\beta^2 - 8\sigma\beta^3 + 180}{11\theta + 6\beta + 8\theta\beta + 14\sigma\beta + 8\sigma\beta^2 + 9}$	$x = \dfrac{1}{2}\dfrac{20\sigma - \theta - \beta - \sigma\beta + 19}{\sigma + 1}$

In what follows, we compare tax rates, emission/production levels, permits prices and welfare under the two institutional settings. Focusing on emissions:

$$q_i - q = \frac{\theta + 2\theta\beta - 2\sigma\beta - 1}{11\theta + 6\beta + 8\theta\beta + 14\sigma\beta + 8\sigma\beta^2 + 9} = -\left(q_j - q\right)$$

Graph 1 represents country i differential emissions between the no harmonization and the harmonization case for a given value of β. As expected, the larger the degree of asymmetry, the larger is the difference in emissions between the two settings. More specifically, country i features more (less) emissions under no harmonization when it is characterized by lower (higher) benefits and a higher (lower) cost parameter related to taxation. Of course, country j shows a symmetric behaviour, as total emissions are given and equal to E.

GRAPH 1

$q_i - q$ (BLACK: β = 2.5; PLAIN GREY β = 9)

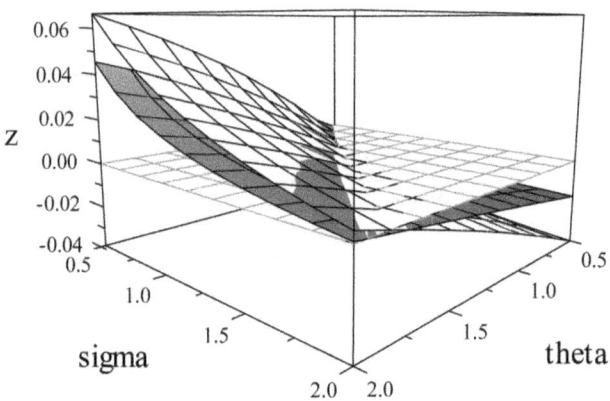

The net buying or selling behaviour of country i under no harmonization is shown in the following expression and in Graph 2:

$$q_i - e_i = \frac{\theta + 2\theta\beta - 2\sigma\beta - 1}{11\theta + 6\beta + 8\theta\beta + 14\sigma\beta + 8\sigma\beta^2 + 9}$$

GRAPH 2

$q_i - e_i$ (BLACK: β = 2.5; PLAIN GREY β = 9)

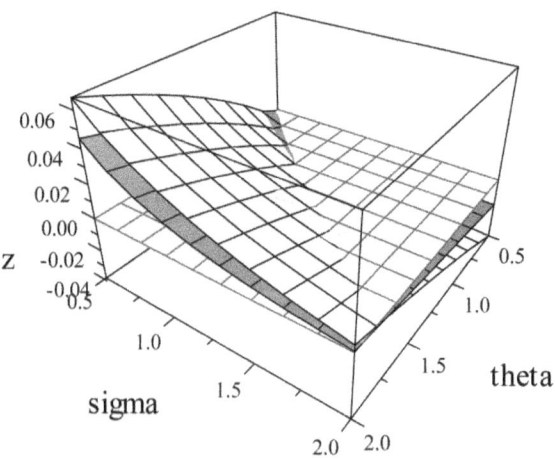

Of course, $q_j - e_j$ follows a graphical pattern which mirrors the one reported in Graph 2, given the equilibrium condition on the permits market. As Graph 2 shows, when country i has a relatively high implementation costs parameter and/or relatively low marginal benefits from public expenditure, it will be a net buyer of permits. Indeed, from the value of q_i in Table 2, we can conclude that, given β and e_i, $\frac{\partial q_i}{\partial \theta} > 0$ and $\frac{\partial q_i}{\partial \sigma} < 0$. As a consequence, for higher θ and lower σ the "total" spillover from expression *(15)* is more likely to be positive and strong. This seems to happen independently of the value for β, although a higher β implies a flatter curve (*i.e.* a lower absolute value of the number of permits bought or sold) as it brings about a lower responsiveness of emissions to changes in the permits price and in the tax rates (see *(3)*). Notice, finally that under our assumed parameter values, $q_i - q = q_i - e_i$.

Turning to tax rates, the comparison implies for country i:

$$t_i - t = \frac{1}{2} \frac{8\sigma^2\beta^2 - 20\theta - 11\theta^2 - 8\sigma\beta - 8\theta^2\beta - 8\theta\sigma\beta^2 - 9}{(\sigma+1)(11\theta + 6\beta + 8\theta\beta + 14\sigma\beta + 8\sigma\beta^2 + 9)}$$

The comparison is reported in Graph 3 for a low ($\beta = 2.5$) and a high ($\beta = 9$) value of β. Notice that independently of β, country i tax rates are lower under no harmonization when country i has relatively small benefits from public expenditure (θ is larger than 1) and/or it features relatively large costs of implementation

GRAPH 3

$t_i - t$ (LEFT: $\beta = 2.5$; RIGHT $\beta = 9$)

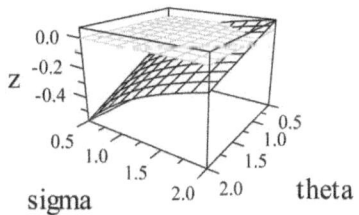 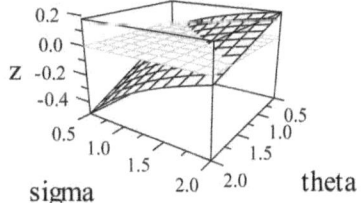

(σ is lower than 1). Also, notice that $t_i - t$ increases with β, *i.e.* a larger β makes it more likely that the no harmonization tax rate in country i is larger than the one arising under harmonization. This is clear if we look at Table 2: while t_i can be easily shown to increase with β (at least when β is such to guarantee that the equilibrium permits price is strictly positive), t is not affected by the value of such parameter.

Turning to country j, the tax rate comparison is reported in the following expression and in Graph 4.

$$t_j - t = -\frac{1}{2}\frac{20\theta + 4\beta + 11\theta^2 - 4\theta\beta + 12\sigma\beta - 8\theta\beta^2 + 8\theta^2\beta + 8\sigma\beta^2 - 4\theta\sigma\beta + 9}{(\sigma+1)(11\theta + 6\beta + 8\theta\beta + 14\sigma\beta + 8\sigma\beta^2 + 9)}$$

GRAPH 4

$t_j - t$ (LEFT: β = 2.5; RIGHT β = 9)

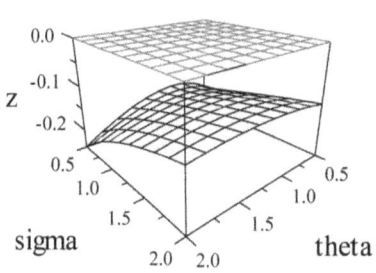 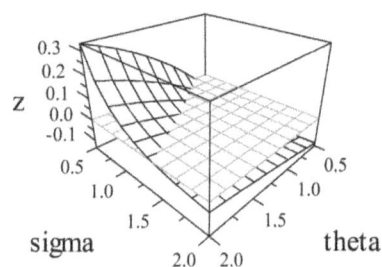

As Graph 4 shows for country j, the mechanics are reversed with respect to country i, but only when β is sufficiently high, weakening the spillover effect as compared to the impact of asymmetries across countries. In Graph 4, for a relatively high β, (right panel) country j will feature a larger equilibrium tax rate under no harmonization if θ is sufficiently large and σ is sufficiently low. On the other hand, for a sufficiently small β, country j, ignoring the resulting (in this case very strong) positive spillover, will set (under no harmonization) a tax rate which is substantially lower than the one resulting under harmonization.

The comparison of permits price depends on tax rates differentials between the two regimes. More specifically, under the assumed parameter values:

$$p-x=\frac{1}{2}\frac{200+2\beta+11\theta^2-4\sigma^2\beta^2-2\theta\beta+10\sigma\beta-4\theta\beta^2+8\theta^2\beta+4\sigma\beta^2+4\theta\sigma\beta^2-2\theta\sigma\beta+9}{(\sigma+1)(11\theta+6\beta+8\theta\beta+14\sigma\beta+8\sigma\beta^2+9)}$$

GRAPH 5

$p - x$ (LEFT: β = 2.5; RIGHT β = 9)

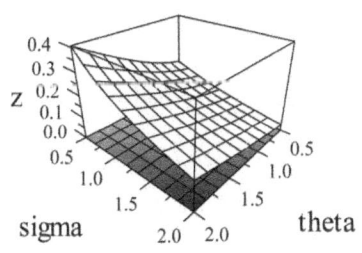
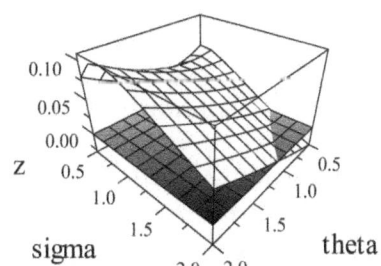

Graph 5 reports the equilibrium permits price differential for varying values of β. For sufficiently low β, the price is higher under no harmonization.

Indeed, for β = 2.5 both countries have been shown to set, in a non harmonized setting, tax rates that are lower than those set under harmonization[8].

In other words, the positive spillover dominates and, given the environmental target, smaller tax rates imply a larger equilibrium price of permits. This does not change unless the spillovers become very weak (*i.e.* β becomes sufficiently high). In such a case what matters is the asymmetry in marginal benefits from public expenditure as well as the differences in implementation costs.

---

[8] Country i would set a larger tax rate in the absence of harmonization only for a relatively high σ and/or a relatively low θ (see Graph 3). This is in any case compensated by the tax rate set by country j, that is always smaller under no harmonization for a sufficiently low β.

Notice that under symmetry ($\theta = \sigma = 1$) the price is in any case larger under no harmonization. Increasing the asymmetry, however, the no harmonization price can turn out to be lower (right panel in Graph 5). This happens, for example, when marginal benefits from public expenditure are much higher in country i and implementation costs of taxation are much lower in the same country. The reason is that, though in such a case the tax rate is lower in country j under no harmonization, the opposite holds with respect to country i; the two effects partially offset each other and, under our assumed parameter values, the "average" tax rate under no harmonization turns out to be higher than the tax rate resulting from harmonization. The equilibrium permits price is therefore larger in the latter case.

Results obtained so far can be summed up as follows.

PROPOSITION 2. *Under assumptions on parameter values guaranteeing strictly positive emissions, tax rates and permits price, a sufficiently large reactivity of emissions to tax rates and permits price (i.e. a low β) makes lower tax rates and a larger equilibrium permits price under no harmonization more likely. As β grows, the equilibrium permits price can turn out to be higher in a harmonized tax setting, if the degree of asymmetry across countries is sufficiently large.*

Our next step is to compare aggregate welfare under the two settings. We will also focus on the impact of harmonization on welfare in single countries. Starting from aggregate welfare, Graph 6 shows $W^h - W^n$ for varying values of β.

As expected, for low values of β the welfare results mimic closely the symmetry case. A low β (top panel in Graph 6) implies that the positive spillover effect dominates any asymmetry, unless such asymmetry is very wide. On the other hand, as β increases a substantial asymmetry across countries might imply a larger aggregate welfare in the absence of harmonization. In the presence of large asymmetries, differentiated taxes allow to exploit the difference across countries in terms of benefits from public expenditures and implementation costs, for example generating revenue

GRAPH 6

$$W^h - W^n \text{ (TOP } \beta = 2.5; \text{ BOTTOM } \beta = 9)$$

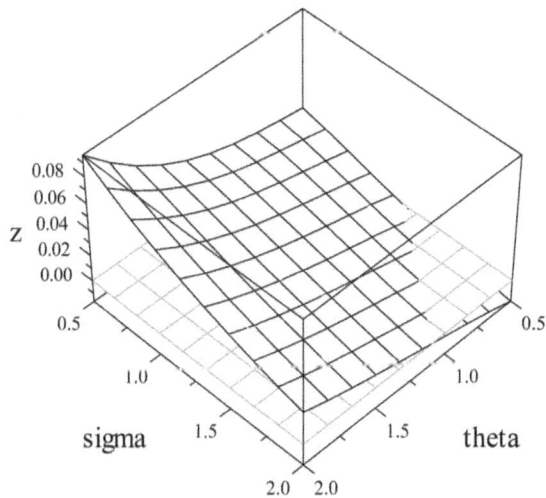

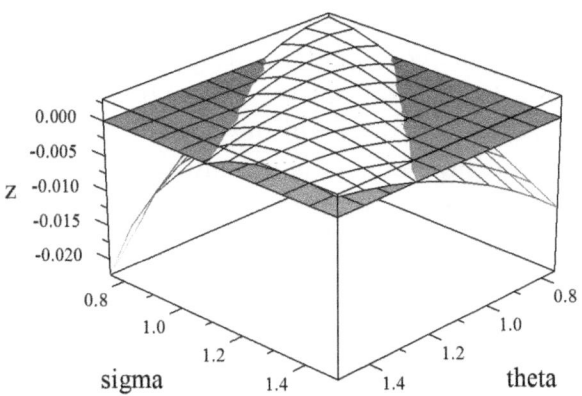

where it is more beneficial and/or where it costs less to do so. As a result, a sufficiently high value of β implies that the spillover effect becomes small, and can be dominated by the impact of asymmetries across countries (bottom panel in Graph 6)[9]. More specifically, when the benefits and costs related to public revenue and taxation are heavily in favour of one country (low σ and high θ or *viceversa*) then harmonization could lead to significant welfare losses.

PROPOSITION 3. *Harmonization could imply a lower welfare than the one arising under differentiated tax rates. This happens for a sufficiently low reactivity of emissions to changes in tax rates and in the permits price and for a sufficiently large degree of asymmetry in favour of one country in terms of benefits and costs of taxation.*

We now turn to welfare in each country. Social welfare differentials in country i and country j, $W_i^h - W_i^n$ and $W_j^h - W_j^n$ respectively, are represented in Graph 7 for relatively low (2.5) and relatively high *(9)* values of β.

When β is sufficiently low, then the left side of Graph 7 shows that country j always gains from harmonization, at least in our assumed parameters range, while country i gains only when the asymmetry in favour of one of the two countries is not too wide. This is the result of the combination of several effects, including benefits and costs from taxation, the net selling position on the permits market as well as the resulting equilibrium price of permits. The impact of these effects is not always straightforward to be interpreted. For example, when σ is small and θ is large country i is, in the absence of harmonization, a net buyer of permits; under the same regime, it sets a smaller tax rate but a larger production level and the price of permits is higher. The compounding of all the above effects leads to a larger country i's welfare in a non harmonized setting.

Welfare impact on the two countries shows a more symmetric behaviour when β increases (on the right of Graph 7). This is

---

[9] The asymmetry across countries might affect the harmonization *vs.* no harmonization comparison also through the differential $\frac{\beta}{2}q_i^2 + \frac{\beta}{2}q_j^2 - \beta q^2$. Though this term affects our analysis quantitatively, it is easily shown not to affect our results qualitatively.

GRAPH 7

TOP $W^h_i - W^n_i$ (LEFT $\beta = 2.5$; RIGHT $\beta = 9$)
BOTTOM $W^h_j - W^n_j$ (LEFT $\beta = 2.5$; RIGHT $\beta = 9$)

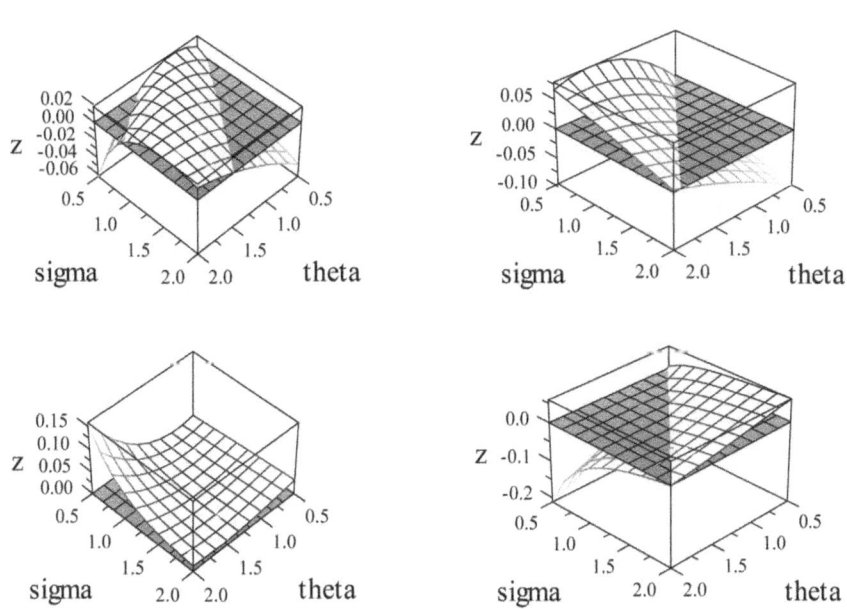

due, at least in part, to the weakening of the spillover effect and to the resulting "symmetric" behaviour of countries in terms of tax setting (see Graphs 3 and 4, right panel).

Interestingly, the comparison of Graphs 6 and 7 shows that there might be cases where harmonization implies improvements in aggregate social welfare but at least one country is worse off. Consider, for example, the case where $\beta = 2.5$, σ is very low and θ is very high. Country i turns out to be worse off under harmonization (Graph 7, top-left) even though aggregate social welfare would be higher (Graph 6, top panel). Indeed, country i is, in these circumstances, quite disadvantaged, featuring relatively high implementation costs and relatively low marginal benefits from taxation. As a result, country i would lose under no harmonization due to a smaller public revenue and to the relatively high price of permits, being country i a net buyer under the assumed parameter values. On the other hand, such losses are likely to be compensated by the relatively low tax rate (and the related implementation costs savings) together with the larger production level, leading country i to prefer a non harmonized framework.

PROPOSITION 4. *Unanimity could not be reached on harmonization even when the latter is aggregate welfare improving.*

The above proposition has relevant policy implications: unless a politically feasible transfer system can be implemented across countries, there might be cases where it would be a social welfare improvement to harmonize tax rates but at least one country would not agree. This might turn out to be crucial in institutional settings such as the EU, where fiscal measures require unanimity to be introduced.

6. - Concluding Remarks

In a simple two countries/two firms setting, we have modelled an environmental regulation problem where a global pollutant is subject to international emissions trading but also to domestic taxation aimed at raising revenue to finance public spending. We have compared two institutional settings, one where emission taxes are harmonized across countries with one where emission taxes are set non cooperatively. Our results show that emission levels, emission tax rates, equilibrium permits price and social welfare might differ in the two settings according to two counteracting forces: a tax related spillover, vehiculated by the international permits market, and asymmetries in the costs and benefits of emission taxation. In particular, the no harmonization case might lead to a larger welfare if sufficiently large differences in favour of one of the two countries exist.

Though our results rely on explicit functional forms and, at least in part, on specific parameter values, we deem our conclusions as theoretically interesting and policy relevant. Under a theoretical point of view, we enrich the received literature by showing that emission tax harmonization could not be the optimal institutional setting when asymmetries in the costs and benefits of taxation are explicitly accounted for. On the policy side, we suggest such asymmetries as a possible rationale behind a lack of consensus over tax rates harmonization.

BIBLIOGRAPHY

AGOSTINI P. - BOTTEON M. - CARRARO C., «A Carbon Tax to Reduce CO_2 Emissions in Europe», *Energy Economics*, no. 14, 1992, pages 279-290.

BOHRINGER C. - KOSCHEL H. - MOSLENER U., «Efficiency Losses from Overlapping Regulation of EU Carbon Emissions», *Journal of Regulatory Economics*, no. 33, 2008, pages 299-317.

EICHNER T. - PETHIG R., «CO_2 Emissions Control with National Emissions Taxes and an International Emissions Trading Scheme», *European Economic Review*, no. 53, 2009, pages 625-635.

GUSDORF F. - HAMMOUDI A., «Heterogeneity, Climate Change, and Stability of International Fiscal Harmonization», *HAL SHS Working Paper* n. 00123293, version 1.9, January 2007.

JOHNSTONE N., *The Use of Tradable Permits in Combination with Other Environmental Policy Instruments*, OECD Environmental Directorate, ENV/EPOC/WP-NEP(2002)28/FINAL, Organisation for Economic Cooperation and Development, 2003.

MARKANDYA A., «Environmental Implications of Non Environmental Policies», in MÄLER K.G. - VINCENT J.R. (eds.), *Handbook of Environmental Economics*, Elsevier B.V, 2005, pages 1354-1401.

MARKUSSEN P. - SVENDSEN G.T., «Industry Lobbying and the Political Economy of GHG Trade in the European Union», *Energy Policy*, no. 33, 2005, pages 245-255.

PEARCE D., *The United Kingdom Climate Change Levy: A Study in Political Economy*, OECD Environment Directorate, COM/ENV/EPOC/CTPA/CFA(2004)66/FINAL, Organisation for Economic Cooperation and Development, 2005.

SORRELL S. - SIJM J., «Carbon Trading in the Policy Mix», *Oxford Review of Economic Policy*, no. 19, 2003, pages 420-437.

SMULDERS S. - VOLLEBERGH H.R.J., «Green Taxes and Administrative Costs: The Case of Carbon Taxation», in CARRARO C. - METCALF G.E. (eds.), *Behavioral and Distributional Effects of Environmental Policy*, National Bureau of Economic Research Inc., 2001, pages 91-130.

ULPH A., «Harmonization and Optimal Environmental Policy in a Federal System with Asymmetric Information», *Journal of Environmental Economics and Management*, no. 39, 2000, pages 224-241.

Index